Communications in Computer and Information Science   1461

Editorial Board Members

Joaquim Filipe
  Polytechnic Institute of Setúbal, Setúbal, Portugal

Ashish Ghosh
  Indian Statistical Institute, Kolkata, India

Raquel Oliveira Prates
  Federal University of Minas Gerais (UFMG), Belo Horizonte, Brazil

Lizhu Zhou
  Tsinghua University, Beijing, China

More information about this series at https://link.springer.com/bookseries/7899

Wai Sze Leung · Marijke Coetzee ·
Duncan Coulter · Deon Cotterrell (Eds.)

ICT Education

50th Annual Conference of the Southern African
Computer Lecturers' Association, SACLA 2021
Johannesburg, South Africa, July 16, 2021
Revised Selected Papers

Editors
Wai Sze Leung
University of Johannesburg
Johannesburg, South Africa

Duncan Coulter
University of Johannesburg
Johannesburg, South Africa

Marijke Coetzee
University of Johannesburg
Johannesburg, South Africa

Deon Cotterrell
University of Johannesburg
Johannesburg, South Africa

ISSN 1865-0929   ISSN 1865-0937 (electronic)
Communications in Computer and Information Science
ISBN 978-3-030-95002-6   ISBN 978-3-030-95003-3 (eBook)
https://doi.org/10.1007/978-3-030-95003-3

© Springer Nature Switzerland AG 2022

This work is subject to copyright. All rights are reserved by the Publisher, whether the whole or part of the material is concerned, specifically the rights of translation, reprinting, reuse of illustrations, recitation, broadcasting, reproduction on microfilms or in any other physical way, and transmission or information storage and retrieval, electronic adaptation, computer software, or by similar or dissimilar methodology now known or hereafter developed.

The use of general descriptive names, registered names, trademarks, service marks, etc. in this publication does not imply, even in the absence of a specific statement, that such names are exempt from the relevant protective laws and regulations and therefore free for general use.

The publisher, the authors and the editors are safe to assume that the advice and information in this book are believed to be true and accurate at the date of publication. Neither the publisher nor the authors or the editors give a warranty, expressed or implied, with respect to the material contained herein or for any errors or omissions that may have been made. The publisher remains neutral with regard to jurisdictional claims in published maps and institutional affiliations.

This Springer imprint is published by the registered company Springer Nature Switzerland AG
The registered company address is: Gewerbestrasse 11, 6330 Cham, Switzerland

Preface

This volume contains selected and revised papers from the 50th Annual Conference of the Southern African Computer Lecturers' Association (SACLA 2021), held on July 16, 2021. The conference was held under the auspices of the Academy for Computer Science and Software Engineering at the University of Johannesburg, South Africa. This is the 5th post-proceedings of SACLA published in Springer's CCIS series. The conference was initially planned to be hosted at the University of Johannesburg, South Africa, but was successfully held online due to the COVID-19 restrictions. Furthermore, due to the open nature of the online conference, university students and other non-presenting delegates were allowed to register at no cost leading to a virtual attendance of more than 80 delegates. SACLA 2021 celebrated 50 years of the continued existence of SACLA with invited talks by Andre Calitz on 50 years of SACLA and by Basie von Solms on the past and future of computing.

The conference theme "Post Pandemic Pedagogy" focused on coordination and resilience-building efforts to ensure education systems' continued sustainability and equity in the COVID-19 pandemic, which has led to unparalleled disruptions in tertiary education. SACLA 2021 invited national and international submissions focussing on practical experiences and successes in computing education. The topic areas addressed included classroom innovations and their impact, novel tools, or novel use of existing tools, and general research in aspects of computing education. In addition, an invited panel discussion with panellists from across the world gave perspectives on online teaching in COVID-19 environments.

The International Program Committee consisted of 22 members, whereby almost half were international (from outside southern Africa). In total, 22 submissions were received from authors around the world, along with one invited submission. A rigorous double-blind refereeing process was followed for all submissions, where each submission was reviewed by at least three Program Committee members. Reviewers were asked to bid on the papers they wanted to review. Then, an automated process allocated papers to each reviewer according to their preferences. After reviews were submitted, the Program Committee accepted 10 full papers based on the review process results. The acceptance rate for the conference was 43%. At the conference, authors of the accepted papers received constructive criticism from the audience to improve the papers before inclusion in this CCIS proceedings.

We want to thank IITPSA for its continued support of SACLA. Many thanks to the members of the International Program Committee, who provided extensive and insightful reviews, and the staff of Springer, who made this CCIS publication possible. We hope that readers will find this volume insightful and look forward to the continuation of the SACLA series in the following years.

November 2021

Marijke Coetzee
Wai Sze Leung
Duncan Coulter
Deon Cotterrell

Organization

General Chair

Wai Sze Leung   Academy of Computer Science and Software Engineering, South Africa

Program Committee Co-chairs

Marijke Coetzee   Academy of Computer Science and Software Engineering, South Africa
Wai Sze Leung   Academy of Computer Science and Software Engineering, South Africa

Local Arrangements and Additional Support

Duncan Coulter   Academy of Computer Science and Software Engineering, South Africa
Deon Cotterrell   Academy of Computer Science and Software Engineering, South Africa

SACLA Publications Chair and CCIS Post-proceedings Co-editor

Marijke Coetzee   Academy of Computer Science and Software Engineering, South Africa

Program Committee

Shirley Atkinson	Plymouth University, UK
Shaun Bangay	Deakin University, Australia
Frans Blauw	University of Johannesburg, South Africa
Alan Chalmers	University of Warwick, UK
Charmain Cilliers	Nelson Mandela University, South Africa
Thomas Clemen	Hamburg University of Applied Sciences, Germany
Sue Conger	University of Dallas, USA
Stephen Flowerday	Rhodes University, South Africa
Patricia Gouws	University of South Africa, South Africa
Jean Greyling	Nelson Mandela University, South Africa
Barry Irwin	Noroff University College, Norway

Janet Liebenberg	North-West University, South Africa
Hugo Lotriet	University of South Africa, South Africa
Jan Mentz	University of South Africa, South Africa
Francois Mouton	Council of Scientific and Industrial Research, India
Liezel Nel	University of the Free State, South Africa
Jacques Ophoff	Abertay University, UK
Rayne Reid	Noroff University College, Norway
Karen Renaud	University of Strathclyde, UK
Aslam Safla	University of Cape Town, South Africa
Ian Sanders	University of South Africa, South Africa
Hussein Suleman	University of Cape Town, South Africa

Contents

Past, Present and Future

The 50 Year History of SACLA and Computer Science Departments
in South Africa 3
Andre P. Calitz

Teaching Innovation

Minimising Tertiary Inter-group Connectedness Over Successive Rounds 27
Andrew Broekman and Linda Marshall

Teaching in a Time of Uncertainty – A Practical Guide 43
Geoffrey Dick

Utilizing Computational Thinking in Programming to Reduce Academic
Dishonesty and Promote Decolonisation 51
Suné van der Linde and Janet Liebenberg

Project-Based Learning Guidelines for IT Higher Education 67
J. T. Janse van Rensburg

Teaching Methods and Strategies

Mapping Computational Thinking Skills to the South African Secondary
School Mathematics Curriculum 85
Karen Bradshaw and Shannon Milne

Common Code Writing Errors Made by Novice Programmers:
Implications for the Teaching of Introductory Programming 102
Mokotsolane Ben Mase and Liezel Nel

Ten Years in the Trenches of a Doubtful Science: An Autoethnographic
Investigation of Five Challenges of Teaching in Information Systems 118
Daniel B. le Roux

Mapping the Problem-Solving Strategies of Novice Programmers
to Polya's Framework: SWOT Analysis as a Bottleneck Identification Tool 132
Pakiso J. Khomokhoana and Liezel Nel

Understanding the Significance of Enterprise Resource Planning Education in Zambia: A Case of an ERP Short Course at University of Zambia 149
  Mampi Lubasi and Lisa F. Seymour

Author Index 165

Past, Present and Future

INTRODUCTION

The 50 Year History of SACLA and Computer Science Departments in South Africa

Andre P. Calitz(✉)

Department of Computing Sciences, Nelson Mandela University, Port Elizabeth, South Africa
Andre.Calitz@Mandela.ac.za

Abstract. The Southern African Computer Lecturers' Association (SACLA) is a formal association of academics involved in education, lecturing and teaching of Computer Science (CS), Information Systems (IS) and related Information Technology (IT) subjects at universities and other Higher Education Institutions throughout Southern Africa. SACLA celebrated 50 years of existence in 2021. The main activity of SACLA is an annual conference, where issues relating to CS, IS and IT education, teaching and research in Southern Africa are presented and discussed. Departments of Computer Science were established at several universities in South Africa during the period 1969 to 1970. SACLA started in 1971 as an initiative of IBM to "Teach the Teacher". The 50-year history of SACLA is presented in this paper, and academics involved in the first SACLA 4-day conferences shared memories and documentation. The paper presents an overview of the 50 conferences hosted, related activities at conferences and the future of SACLA.

Keywords: SACLA History · TECLA · CSLA · Computer Science Departments

1 South African Computer Science Departments

Computer Science as a discipline began in the 1950s and early 1960s. The world's first Computer Science degree programme, namely The Cambridge Diploma in Computer Science, began at the University of Cambridge Computer Laboratory in the United Kingdom in 1953. The term Computer Science was created by George Forsythe, a numerical analyst. The first Department of Computer Science was formed at Purdue University in 1962 [16]. The first person to receive a PhD from a Computer Science department was Richard Wexelblat, at the University of Pennsylvania in December 1965.

The first FORTRAN compiler was developed in April 1957. LISP, a list-processing language for artificial intelligence programming, was invented in 1958 [16]. New programming languages, such as BASIC was invented in the 1960s, and during this time, we saw the rise of automata theory and the theory of formal languages. A rigorous mathematical basis for the analysis of algorithms began with the work of Donald E. Knuth, author of a 3-volume treatise entitled The Art of Computer Programming [16]. The 1970s saw advances in databases' theory, and Edgar F. Codd made significant contributions to relational databases.

In South Africa, several Computer Science departments were established during the late 1960s and early 1970s. At the University of Stellenbosch, the Department of Mathematics in 1969 introduced a one year course in Computer Science and in 1970, the Department of Computer Science was established with Professor George Murray as its first head of department (HOD) [6]. Professor D. G. Parkyn was head of the Applied Mathematics department at the University of Cape Town (UCT) until 1970 when he became acting head of the newly formed Computer Science Department. The computer hardware in 1970 at UCT consisted of an IBM 1130 with 16k memory, two 512k disk drives and a graph plotter. The first Computer Science I course at UCT included modules on machine organisation, Assembler and FORTRAN programming techniques and Information Structures and applications. The Computer Science II in 1971 included modules on Systems analysis, COBOL, ALGOL, Operating systems, Compiler construction and Formal languages. A lecturer at Glasgow University, Professor Ken MacGregor was appointed as Senior Lecturer at UCT in 1973. Professor Sonia Berman joined the department in 1982, and Professor Pieter Kritzinger joined in 1986 and became HOD in 1988 [17].

Professor A.P.J. der Walt was appointed in 1970 as Chairman of the Department of Computer Science at the Rand Afrikaans University (RAU) (Fig. 1). The Rand Afrikaans University, today called the University of Johannesburg (UJ), installed an IBM 1130 in October 1969, shown in Fig. 1.

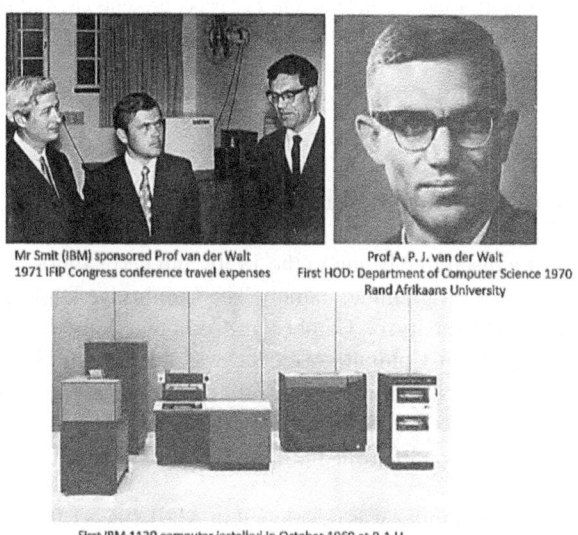

Mr Smit (IBM) sponsored Prof van der Walt
1971 IFIP Congress conference travel expenses

Prof A. P. J. van der Walt
First HOD: Department of Computer Science 1970
Rand Afrikaans University

First IBM 1130 computer installed in October 1969 at R.A.U.

Fig. 1. RAU first HOD and the IBM 1130.

In 1971, IBM sponsored Professor van der Walt's travelling expenses to attend two international conferences related to Computer Science. The first was the "International Symposium on the Theory of Machines and Computations", which took place in Haifa, Israel and the second, the "IFIP Congress 71", was held in Ljubljana, Yugoslavia. It was the first time a representative of a South African university participated in one of three annual computer conferences [5].

In 1967, the University of South Africa (UNISA) was the first South African university to offer Human-Computer Interaction (HCI) courses at the under-graduate and post-graduate levels. It instituted a course to train students in the field of Computer Science. The Department of Computer Science in 1970 included Prof R.H. Venter (HOD), Prof C.H. Bornman and Mr G. de V. de Kock and Dr G.J. Joubert as Senior Lecturers. In 1970, UNISA received a grant of R40,000 from International Computers South Africa PTY. Ltd. (ICL) to establish a fellowship scheme to alleviate the country's severe shortage of computer scientists [9]. During 1973, the number of first-year Computer Science students grew to 1,821. At his inaugural lecture in 1975, Prof Chris Borman stated that "The computing education provided by the universities is inadequate to the needs of industry. The Computer Science graduates had learned too much about Numerical Methods Theories of computation, and they were fluent in every computer language except English. However, they did not know how to apply the concepts they had learned to the typical commercial environment" [9]. In 1980 a new course, COBOL (Commercial Programming Course), was introduced to produce professional, well-trained programmers ready to enter the job market.

Rhodes University (RU) acquired its first computer in 1966, and Computer Science was first introduced as a major subject under the auspices of the Department of Applied Mathematics in 1970, and a separate Computer Science department was established in 1980 [20]. Professor Pat Terry was the first HOD for the Computer Science department (Fig. 2).

At the University of the Free State (UFS), Computer Science was first presented in 1973 as a course in the Department of Statistics. In 1980, The Department of Computer Science and Information Systems was established with Professor T.H.C. Smith appointed as HOD (Fig. 2). In 1983, the HOD was Professor Theo Mc Donald until 1987 when Professor Hans Messerschmidt was appointed HOD [7].

In 1969, after completing his PhD at the University of California, Berkeley, Professor Roelf van den Heever (Fig. 2) returned to South Africa to establish what has grown into the Department of Computer Science at the University of Pretoria (UP) [12]. In those days, Computer Science was presented under the auspices of the Department of Statistics. However, in 1975 it broke away to form an independent entity within the Faculty of Natural and Agricultural Sciences. Professor van der Heever was HOD from 1969 to 1997. The first graduates received their BSc Honours degrees in Computer Science in 1971 [10].

The North-West University (NWU), formerly known as the Potchefstroom University for Christian Higher Education, installed an IBM 1130 in September 1967 in the Computer Center. A second IBM 1130 was purchased in 1970/1971, and the Department of Computing Science was established. The first Computer Science courses at the second-year level were presented in 1971, with Professor Hannes de Beer appointed as HOD. In 1973 a Computer Science Honours degree was introduced, followed by MSc and PhD degrees in 1974 [18].

The North-West University (NWU), formerly known as the Potchefstroom University for Christian Higher Education, installed an IBM 1130 in September 1967 in the

Prof Theuns Smith
First HOD: Department of Computer Science 1980
University of the Free State

Prof Gerrit Wiechers
First HOD: Department of Computer Science 1970
University of Port Elizabeth

Prof Pat Terry
First HOD: Department of Computer Science 1980
Rhodes University

Prof Roelf van den Heever
First HOD: Department of Computer Science 1975
University of Pretoria

Fig. 2. Departments of Computer Science HODs

Computer Center. A second IBM 1130 was purchased in 1970/1971, and the Department of Computing Science was established. The first Computer Science courses at the second-year level were presented in 1971, with Professor Hannes de Beer appointed as HOD. In 1973 a Computer Science Honours degree was introduced, followed by MSc and PhD degrees in 1974 [18].

The first call for establishing a Computer Science department at the University of Port Elizabeth (UPE), now known as the Nelson Mandela University (NMU), was made in 1968 by Professor Gerrit Wiechers, Head of the Department of Applied Mathematics, in a speech during a Senate meeting (Fig. 2). In 1969, the Council of UPE agreed to establish a Computer Center; after that, a Computer Science Department was approved for subsidy purposes. A contract was signed for the lease of an ICL 1901A mainframe computer at R1200 per month. This computer was used for administrative, research and teaching purposes [13]. The first HOD of Computer Science was Professor Gerrit Wiechers. He was the only lecturer, teaching 18 first-year students and two honours students in 1970. In 1975, Professor Gideon de Kock was appointed as HOD and remained for 30 years.

2 SACLA First Meetings

International seminars on the teaching of Computer Science were an initiative that arose from an initiative by the then Director of the Computing Laboratory at Newcastle University [11], Professor Ewan Page, who had attended a similar, but once-off event organised by IBM in Paris in 1968. These annual four-day seminars continued for 34-years at the University of Newcastle [11]. They brought together an invited audience of senior UK and European computing academics to hear a series of presentations from distinguished international speakers. Each seminar concentrated on a particular theme (Fig. 4), usually a major Computer Science research domain. A small group of speakers each gave one or more lectures reviewing their own and other contributions to the development of the Computer Science curricula. The 4-day seminars at the Newcastle University, 1968–1971 included speakers, such as Niklaus Wirth, Ed Coffman, Don Knuth and Edsger Dijkstra. The topics covered by different presenters mainly focused on the teaching of Computer Science (Fig. 4). Links to the seminar topics, presenters and the papers for the period 1968 to 2001 are provided on the University of Newcastle website [11].

The Southern African Computer Lecturers' Association (SACLA) [14] was established 50 years ago by academics and IBM, an industry computer mainframe supplier. Initially, the association was called the Computer Science Lecturers' Association (CSLA). The CSLA documents published in 1982 and 1983 (Fig. 7) displayed the CSLA name; however, in 1984, the conference was called SACLA.

Fig. 3. IBM 1130 mainframe and Ewan Page, guest speaker and lecturer 1970–1973.

The South African academics involved with the first CSLA meetings were Roelf van den Heever [19] and the late Dewald Roode. CSLA was initially an informal association of academics involved in education, lecturing, and teaching Computer Science. The main activity of CSLA was an annual conference, which started in 1973.

In the 1970s, IBM in South Africa started an initiative called "Teach the Teachers", where they brought out lecturers from the UK to give a week's classes to everyone who was embarking at that time on teaching Computer Science. In 1970, RAU, NWU and UCT were using an IBM 1130 mainframe computer with punch cards (Fig. 3). The first CSLA Conferences (4-day meetings) focused on Computer Science curricula, presentation methods and practical work [2].

The first three CSLA seminars were held in the Engineering building of Stellenbosch University, where the Department of Computer Science is now housed [2]. The CSLA seminars followed a similar format as the "Teach the Teachers" initiative presented at Newcastle University and was sponsored by IBM. The four-day conference consisted mainly of lectures presented by Professor Ewan Page and other academics from the UK, including Professor David Barron from Southampton University. He spoke on programming languages PL360, a system programming language designed by Nicklaus Wirth in the 1960s [2]. The conference delegates were "everyone who was embarking at that time on teaching Computer Science" [2].

Newcastle International Seminars - On The Teaching Of Computing Science: 1968—2001				
1968 — Computer Science		1970 — The Teaching of Programming at University Level	
D. T. Ross	Data Structure and Storage Management	E. W. Dijkstra	The Art of Programming
N. Wirth	Teaching Compiler Design — Outline of a Proposed Course	D. E. Knuth	The Analysis of Algorithms
D. N. Freeman	A Study of Stack Architecture in Control Program Design	K. E. Iverson	The Use of APL in Teaching
J. du Masle	Software Teaching and Research at the University of Grenoble	M. A. Jackson	The Construction of Algorithms in a Commercial Programming Environment
E. S. Page	Constraints on Teaching in Computing Science	W. C. Lynch	The Creation of Systems Programmers
1969 — On the Teaching of the Design of Large Software Systems		1975 — Computers and the Educated Individual	
B. W. Arden	Multi-Processing Systems	W. A. Clark	The Basis of Present Computer Design
A. B. Cleaver	Systems Evaluation	P. C. Goldberg	The Future of Programming for Non-Programmers
C. J. Bell	Information Systems	F. J. M. Laver	Computers and Society
E. G. Coffman Jr	Formalism in Computer System Design — Models of Parallelism and Concurrency	P. Naur	An Adaptable course of Elementary, University Level, Computer Science
K. I. McKenzie	The Problem of Debugging the Large On-Line System	A. W. Holt	Formal Methods in System Analysis

Fig. 4. Speakers and topics presented 1968–1975.

Figure 5, supplied by Professor Judith Bishop [2], shows the 1972 CSLA conference representatives. First, seated in front are Professor Page and other IBM representatives. Next, standing sideways, Professor Judith Bishop and Professor Basie von Solms are to the right in the picture. Forty-eight delegates attended the 1972 conference.

The first 15 years of CSLA conference organisers are shown in Fig. 6. Professor Page was the guest speaker at the first three seminars that focused on Computer Science curricula, teaching programming and computer design. The sponsors of the first SACLA conferences were mainly IBM; however, in later years, other sponsors were Burroughs, ICL, NCR and Univac (Fig. 6). The late Professor Dewald Rhoode and Professor Roelf van den Heever from the University of Pretoria were responsible for organising the conferences at Golden Gate from 1974 to 1976 [8].

The first 15 years of CSLA conference organisers are shown in Fig. 6. Professor Page was the guest speaker at the first three seminars that focused on Computer Science curricula, teaching programming and computer design. The sponsors of the first SACLA conferences were mainly IBM; however, in later years, other sponsors were Burroughs, ICL, NCR and Univac (Fig. 6). The late Professor Dewald Rhoode and Professor Roelf van den Heever from the University of Pretoria were responsible for organising the conferences at Golden Gate from 1974 to 1976 [8].

Professor Roelf van den Heever [19] indicated that many SA universities started Computer Science programs in the early 1970s. Computer Science was a brand-new

Fig. 5. Attendees at CSLA meeting in 1972 (Source: Judith Bishop).

HISTORY OF SACLA

Year	University Chair(S)	Guest Speaker	Sponsor(S)	Venue
1971		Ewan Page	IBM	Jan Smuts
1972		Ewan Page	IBM	Senator, Bellville
1972		Ewan Page	IBM	
1974	R Venter		IBM	Golden Gate
	D Roode			
1975	R Venter		Burroughs	Golden Gate
	D Roode			
1976	John Shochot		Univac	Golden Gate
1977	John Shochot	David Barron	ICL	Hunter's Rest, Rustenburg
1978	Ken MacGregor	B Mittman	CDC	Hunter's Rest, Rustenburg
1979	Judy Bishop		Univac	Golden Gate
1980	Judy Bishop		Univac	Golden Gate
1981			Burroughs	Gordon's Bay
1982	Dennis Rordan	Henry Lucas	ICL	Wigwam, Rustenburg
1983	Edwin Anderson	C Wogrin	CDC	Transkei Holiday Inn
1984	Alan Sartori-Angus		Univac	Hluhluwe
1985	Stef Postma		NCR	Kimberley

Photos: Ewan Page; Don Cowan IBM; Henry Lucas Prof IS: MIT; Dewald Roode: UP; Ken MacGregor: UCT; Denis Riordan: Rhodes

Fig. 6. First SACLA conferences and organisers 1970–1985.

field, as witnessed by introducing the first model curriculum, called "CURRICULUM 68 Recommendations for Academic Programs in Computer Science". This was a report of the ACM curriculum committee on Computer Science. The workshops organised by IBM during 1970, 1971, and the start of 1973 were well received and significant interest in cooperation and collaboration amongst SA computer scientists was a natural consequence. In 1973, Roelf van den Heever approached IBM for financial assistance for the Golden Gate gathering of 26 interested CS leaders from all SA universities. Roelf got in touch with Dewald Roode (RAU) and Gerrit Wiechers (UNISA) to form a coordinating committee [19].

The final session at the 1974 Golden Gate conference was under the chairmanship of Derek Henderson [19]. The purpose was to plan further activities. The following decisions were made:

- Dewald Roode, Roelf van den Heever and Gerrit Wiechers were elected to form an Executive Committee;
- George Murray (US) was given the task of determining the minimum requirements for Computer Science education and verifying that these agreed with what the outside world was expecting;
- Rolf Braae (Rhodes) was tasked with putting forward proposals for new topics that should be covered in the Computer Science curriculum; and
- Gerrit Wiechers (UNISA) was tasked with taking care of the general duties of liaison, including the investigation of extension courses, liaison with colleges of advanced technology and liaison with the Department of Education.

Here are some quotes from notes that were taken during the concluding session [19]:

- "Will teaching staff, as well as heads of Computer Centres at Universities, be eligible for membership of our group?"
- "The executive committee will give more details on who will be eligible for membership. We ought to give attention to giving the group a name but must be careful so that nobody would be excluded from future enhancement of the group."

The meeting decided to postpone a decision on a name for the Group until the third sub-committee had met with the Computer Society. For the time being, the Group will go under the name of "Work Group for Computer Education" – Afrikaans: "Werkgroep vir Rekenaaropleiding". Concluding remarks by chairman Henderson: "I should like very much to express on behalf of all of us our enormous appreciation of IBM's willingness to sponsor this conference we have had at Golden Gate. Secondly, I should like to thank Professors Wiechers, Roode and van den Heever for their initiative, enthusiasm and work. We owe an enormous amount of gratitude to these people."

The Chair of the 1978 CSLA conference was the late Professor Ken MacGregor from the University of Cape Town (Fig. 6). In 1979, Don Cowan was at Waterloo University, Canada, an expert on WatFOR – a fast Fortran compiler for IBMs. Professor Bishop states, "He was a great speaker and helped many students after that. However, the topic was slightly controversial as Wits was using Pascal!" [2]. In 1982 Professor Henry Lucas, a Professor in Information Systems at MIT was the guest speaker. The author contacted him via email. However, he indicated that after 50 years, he could not recall the topic. He remembered that Jonathan Miller from the UCT Business School invited him.

The conference chair in 1982 was Professor Denis Riordan from Rhodes University, the HOD from 1980 to 1981, who later left to go to Canada in the 1990s. Professor Alan Sartori-Angus, conference chair in 1984, was at Rhodes for several years in the 1990s [3].

3 CSLA News Publications 1982 and 1983

Edwin Anderson [1], the Chairman of the Computer Science Lecturers' Association (CSLA) in 1983, provided copies of the CSLA News published from 1982 to 1983, as shown in Fig. 7. The newsletters contained conference news, news relating to different departments, and the results of an annual survey of mainframe equipment used in one issue. In addition, student access to micro-computers and other facilities, such as scanners, etc., was included. The publication also included the conference programme and a list of all universities and technikons, including all personnel working in Computer Science and other related departments. The November 1982 edition focused on curriculum design and listed each university and the programmes offered. The March 1983 issue focused on the conference and included a list of the prescribed books used for different courses by universities and technikons.

The newsletter also contained information relating to departments; for example, the September 1982 letter indicated that Professor Judy Bishop from WITS Computer Science department reported an R100 000 donation to purchase a VAX-370 system with 12 terminals for use by personnel and senior students. Professor Trevor Crossman, WITS Business Information Systems Department, was promoted, and WITS changed their Business Information Systems course from two years to three years. Professor Theuns Smith returned after seven months of database research at the University of Pennsylvania; Professor Sartori-Angus from Rhodes joined the University of Natal. These four publications have been professionally scanned and placed on the SACLA website [14].

The newsletter provided a list of SACLA members, academics working in the different departments at various universities from 1983 to 1984. One hundred delegates attended the 1983 conference. Professor Sonja Bergman, conference chair in 1986, is listed as a Junior Lecturer at UCT, together with the late Professor Ken MacGregor and Professor Keith Mattison, then a Senior Lecturer. Staff were listed from universities, such as Fort Hare, Natal, University of the North, Potchefstroom, UPE, RAU, US, WITS and Rhodes. Professor Philip Machanick was a junior lecturer at WITS.

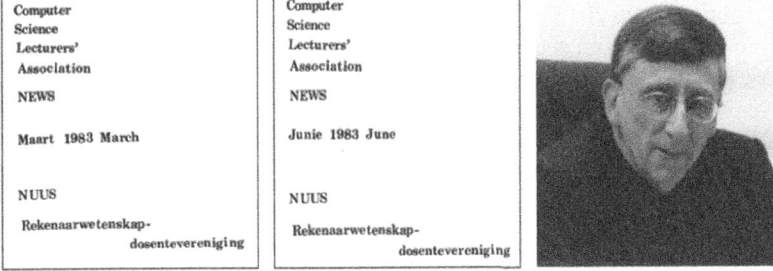

Fig. 7. CSLA 1983 news. Picture of Edwin Anderson.

4 TECLA and HEICTA

The CSLA March 1983 newsletter indicated that two topics would be discussed at the 1983 annual conference: each department's approach to research and the future development of Computer Science and the role of universities and technikons. The conference programme indicated that each department had to send a representative to present the department's research areas. Professor Roelf van den Heever and Professor Denis Riordan were the two speakers to summarise the research conducted in South Africa. The conference programme also indicated that Professor Dewald Roode (UP) would provide an overview of the role of Computer Science at universities, and Mr Hennie Groenewald (Pretoria Technikon) would lead the discussion from a Technikon viewpoint.

At the SACLA conference AGM in 1984, held in Hluhluwe, Natal, South Africa, a heated debate between academics from universities offering CS degree programmes and technikons offering IT diploma programmes. Professor Rossouw von Solms [23] indicated that the technikons at that time only had observer status at SACLA without any voting rights. This discussion at the AGM resulted in the academic members from the Technikons in South Africa being disassociated from SACLA.

Professor Rossouw von Solms initiated the formation of an independent body with similar aims to SACLA, namely the Technikon Computer Lecturers Association (TECLA), in 1989, and he was the first President of TECLA [4, 23]. The primary purpose of TECLA was to have an annual conference where the Technikon academics discussed the various IT diploma programmes and curriculum content in South Africa. However, TECLA ceased to exist when the South African Government introduced new university structures in 2005 and amalgamated six technikons and specific universities, forming six comprehensive universities and creating eight Universities of Technology.

A new body, Higher Education Information and Communication Technology Association (HEICTA), was established in 2006 and was also involved with industry liaison and diploma programme curriculum topics. For several years, it provided collaboration among IT departments offering IT diplomas and liaised with the industry. However, HEICTA gradually ceased to exist. Professor Bennett Alexander of CPUT was the president of HEICTA. Documentation on the functioning of HEICTA was found by the authors dating back to 2013 [4]. After then, HEICTA members have not met annually, and currently, all academics from all universities involved in IT education in Southern Africa are members of SACLA. All institutions offering IT diplomas are now responsible for registering their qualifications with the South African Qualifications Authority (SAQA).

The minutes of the SACLA AGM on 28 June 2002, organised by Rhodes University, indicated that "Kevin Johnston proposed that the current SACLA committee contact TECLA to initiate a merger. It was also proposed that the 2003 SACLA committee invite TECLA members to attend the conference in June 2003". The minutes of the AGM held at Pilanesberg National Park on 1 July 2003, Chaired by Tom Addison from WITS University, again indicated under point 9 - Technikon Considerations that "The idea of inviting technikons (and other institutions) needs to be explored. The changing status of technikons might negate the issue. After some discussion on the advantages and disadvantages of inviting technikons, it was decided that there would be no specific drive to invite them, but that there would be no problem with their attending. The committee should look at universities more generally/holistically". A discussion to invite TECLA members were again included at the 2006 AGM. A resolution was adopted at the 2014 AGM that Technikon staff were included in the membership body.

5 SACLA Hosting Universities 1980 to Present

The Chair of the 1985 SACLA conference, held in Kimberley, South Africa, was Professor Stef Postma. Professor Sonja Bergman from the Department of Computer Science at the University of Cape Town was the conference chair for the 1986 conference held at the Amatola Sun, close to Bisho in the Eastern Cape, shown in Fig. 8. The 1988 SACLA conference chair was Professor Peter Warren from the University of Port Elizabeth, now the Nelson Mandela University and the 1992 conference chair was Professor Judith Bishop. The conference was held in Rustenburg, South Africa, and this was the third time Judith acted as conference chair. The sponsors during these years included ISM (IBM in SA), Unisys and Persetel.

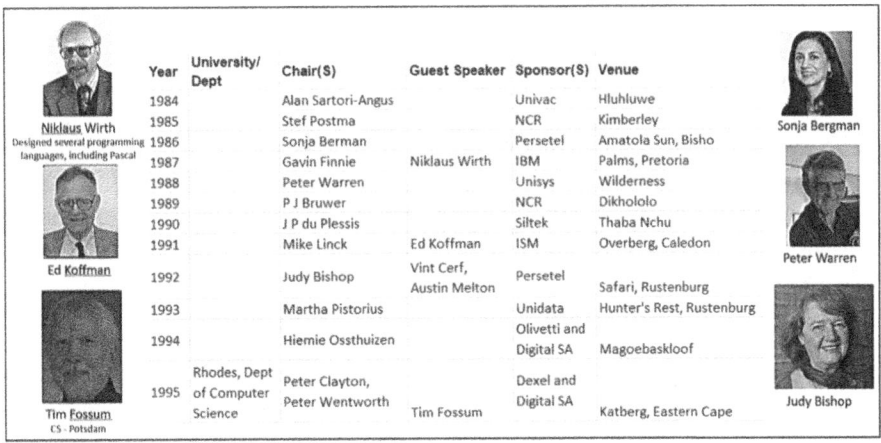

Fig. 8. SACLA hosting 1986 to 1995

In 1987, the guest speaker was Niklaus Wirth, a Swiss computer scientist. He has designed several programming languages, including Algol and Pascal. He also worked in the field of Software Engineering, and in 1984, he received the Association for Computing Machinery (ACM) Turing Award for the development of various computer programming languages. In 1991, the guest speaker was Ed Koffman from the Mathematics Research Center of Bell Laboratories. Ed was the author of various Computer Science textbooks and received the SIGCSE Outstanding Contribution Award in 2009, specifically for "helping to shape Computer Science education". Finally, Tim Fossum from the Computer Science Department at the University of Wisconsin was the keynote speaker in 1995 (Fig. 8).

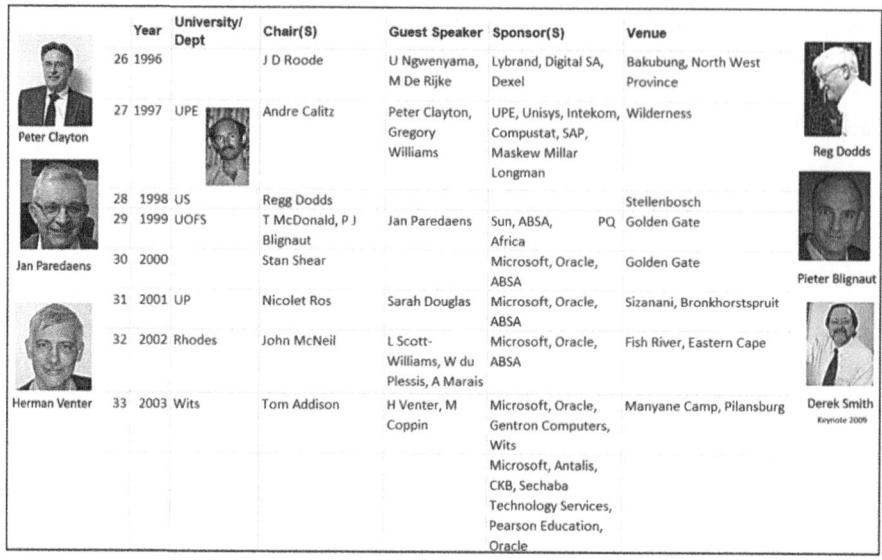

Fig. 9. SACLA hosting 1996 to 2003.

The 1996 conference was chaired by the late Professor Dewald Rhoode, his third time as conference chair (Fig. 9). The author's first time as conference chair was in 1997 when SACLA 1997 was held in the Wilderness. Professor Peter Clayton from Rhodes was the guest speaker, and Professor Reg Dodds chaired the 1998 conference. Professor Pieter Blignaut was conference chair in 1999, and the guest speaker was Professor Jan Paredaens from the University of Antwerp in Belgium. He was the author of various books on programming languages, data structures and relational databases.

The first AGM minutes stored on the SACLA website (Fig. 10) were the Minutes of the 2001 AGM. Professor Derrick Kourie from UP chaired the meeting. The financial report indicated that R160 000 was obtained in sponsorships from Microsoft, Oracle and ABSA. The treasurer for the conference was Dr Linda Marshall from the Department of Computer Science at UP.

The 2002 conference was held at the Fish River Sun and was organised by RU. One hundred and ten delegates attended the conference. Professor Herman Venter, formally an academic in the Department of CS&IS at UPE, was the guest speaker at the SACLA 2003 conference. Tim Addison was the conference chair, and the conference took place in Pilanesberg nature reserve, South Africa (Fig. 9).

UKZN hosted the 2004 conference. At the 2005 conference, the guest speaker at the Mowana Safari Lodge in Botswana was Professor Frank Youngman, who specialised in adult education. The AGM minutes dated 5 July 2005 indicated that 34 delegates attended the AGM. The AGM minutes indicated that the Technikons do not exist anymore and "SACLA has removed all impediments in the way of Technikons so that Technikon academics can submit any number of their papers". It was further suggested that a body be formed to represent the interest of academia.

Southern African Computer Lecturers' Association
SACLA 2001

Minutes of the Annual General Meeting
held at 14:30 on 29 June 2001 at Sizanani

1. Attendance

Roundabout fourty delegates attended the AGM.

2. Welcome

Prof Derrick Kourie from the Department of Computer Science, welcomed everyone present on behalf of the organising committee.

3. Minutes of SACLA 2000 AGM

The minutes of the AGM for SACLA 2000 were distributed and adopted.

4. Chairman's report

Prof Kourie thanked the organising committee for their hard work and the sponsors, Microsoft, Oracle and ABSA, for their generous contributions.

A question was raised with regard to sponsorships. The organising committee did not need to ask for sponsors, since they were contacted by them well in advance. During December 2000, Microsoft had already committed to support SACLA 2001.

5. Financial report

Income	
Sponsorship	R 160,000.00
From SACLA 2000	R 15,000.00
Conference Fees	R 10,420.00
Other	R 1,000.00
TOTAL	**R 186,420.00**
Expenditure	
Accomodation, meals and dinner	R 55,968.81
Printed Matter	R 10,598.89
Gifts, wine etc.	R 21,100.54
Entertainment	R 2,500.00
Transport	R 1,400.00
Support Staff	R 23,000.00
Donation to Sizanani Trust	R 10,000.00
Other	R 4,950.00
To SACLA 2002	R 15,000.00
TOTAL	**R 144,518.24**
Difference (Income - Expenditure)	**R 41,901.76**
Current Balance	**R 41,996.00**

(No interest or service fees have been taken into account)

Prepared by Linda Marshall - 21/06/2002

6. General

All attendees supported a motion to especially thank the UP team for a very well-orgnised and successful conference.

7. 2002 Conference and Organising Committee

Rhodes University will host the SACLA 2002 conference but John Roberts will first confirm this with his colleagues since he is the only Rhodes representative at SACLA 2001. The IS department of Wits will serve as "backup host".

8. Closing

Prof Kourie wished everybody a safe journey home and declared the meeting closed.

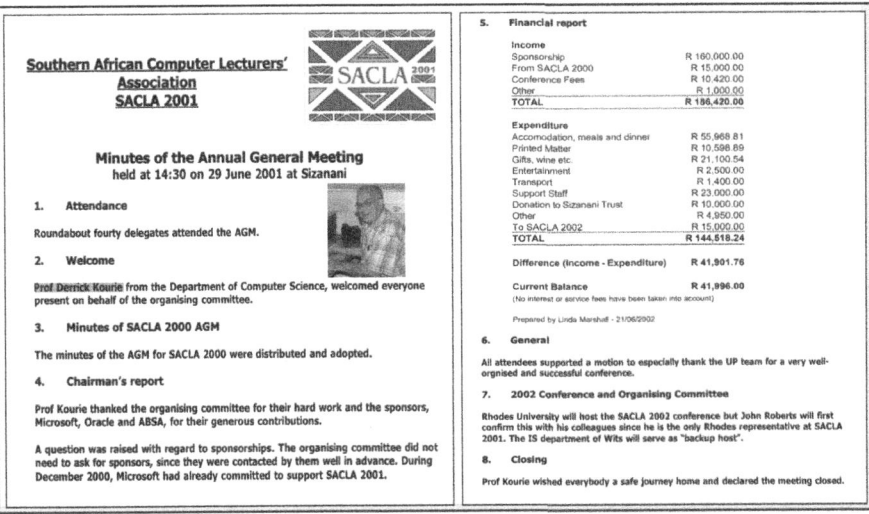

Fig. 10. SACLA 2000 AGM minutes with Chair Professor Derrick Kourie.

	Year	University/ Dept	Chair(S)	Guest Speaker	Sponsor(S)	Venue
34	2004	UKZN	Manoj Maharaj	Simon Turner	Hewlitt-Parkard, Wiley Publishers, Eclipse Networks, Telkom, ETA Audiovisual	Rob Roy Hotel, Botha's Hil
35	2005	Univ of Botswana	Sampson Asare	Prof F Youngman	Univ of Botswana, Botswana dept of IT, Ministry of Comm, Science, Tech, Air Botswana, ICL, Botswana Technology Centre, Microsoft, DCDM Consulting, Botswana dept of Tourism	Kasane, Botswana
36	2006	UCT	Jeann-Paul van Belle	Viola Manuel	BULL, Microsoft, Oracle, NBN	Somerset West
37	2007	Univ Zimbabwe	Bayeza Mangena	Michael Luck	Microsoft	Elephant Hills, Zimbabwe
38	2008	UNISA				Berg-en-Dal
39	2009	Rhodes IS and CS	John MacNeil	Derek Smith	ABSA, Microsoft, Cengage Learning	Mpekweni Beach Resort
40	2010	UP IS	Pieter Joubert	Heikki Topi	Oracle, Microsoft and Cengage Learning	Zebra Country Lodge
41	2011	UKZN	I Govender			Fairmont Zimbali Resort
42	2012	UFS IS and CS	Eduan Kotze	Johannes Cronje	Microsoft, Dell, Oracle, UFS	Black Mountain Leisure & Conference centre
43	2013	Univ of Botswana				*cancelled*

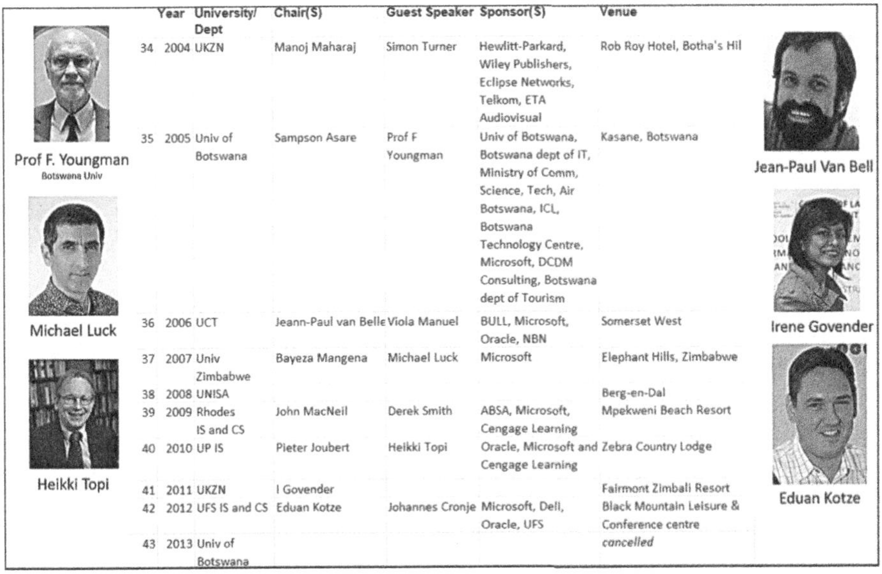

Fig. 11. SACLA hosting 2004 to 2013.

UCT and Professor Jean-Paul van Belle hosted the 2006 conference as the conference chair (Fig. 11). The minutes of the 2006 AGM included the agenda item "Establish a more permanent/formal SACLA body". Tom Addison initiated this discussion by highlighting the problems experienced with the current SACLA modus operandi. He mentioned that he had been observing problems with organising SACLA conferences over the last few years. Most of these were due to the lack of continuity when a new host university assumed the role of Chairman and executive committee. The concerns raised included "Delays of 11 months in receiving the previous year's minutes and seed money", ignorance on hosting a conference, no forward plan for future hosting and "Every new host has to open a new bank account".

Tom Addison proposed a three-person executive committee, Chair of the incoming host, a nominee of the outgoing committee and one other member. He informed the meeting that the constitution of SACLA stipulates a 3-person Exco, but this has never been implemented. It was agreed that an Exco was formed at the meeting, and the following members were elected unanimously: Tom Addison, Jean-Paul Van Belle (SACLA 2006) and Mthulisi Velimpini (SACLA 2007). A competition to create a standardised SACLA logo was also proposed, and a permanent SACLA bank account was created.

The 2007 conference at the Victoria Falls, Zimbabwe, was attended by 9 South African delegates, namely I. Brown, V. Midze, R. von Solms, B. von Solms, J. Greyling, V. Pieterse, L. Goosen, W. Nel and J. van Biljon. The guest speaker was Professor. Michael Luck, a computer scientist, based at the Department of Computer Science, King's College London, England. His research includes intelligent agents and multi-agent systems, and he has over 300 publications.

The guest speaker at the 2009 conference held at the Mpekweni Beach Resort in the Eastern Cape was Professor Derek Smith from UCT. John McNeill was conference chair, and 50 delegates attended from 14 institutions. The theme of the conference was "Connecting Students to the Future". Thirty-three members attended the AGM, and the discussions included the publication of the conference proceedings in a journal. It was agreed that the editor of SACJ, Lucas Venter, would be approached.

The UP Department of Informatics hosted the SACLA 2010 conference from 3 to 5 June 2010 at Zebra Country Lodge just north of Pretoria, and the theme of the conference was "Scoring IT Education Goals in 2010". The keynote speaker was Professor Heikki Topi from Bentley University, who spoke on the newly published IS2010 Changing the Course for Undergraduate IS Model Curricula.

Professor Irene Govender chaired the 2011 SACLA conference from the University of Kwazulu Natal, and Professor Eduan Kotze (UFS) chaired the 2012 conference. The 2012 conference was held at the Black Mountain Leisure and Conference Hotel. Thirty-two delegates attended the AGM, and it was recommended that Botswana University host the 2013 conference. However, the sponsors indicated that they could not sponsor SACLA if it is situated outside the RSA borders, and Botswana will have to find their sponsors (Fig. 11).

The 2013 SACLA conference was cancelled; thus, no official handover was made to a 2014 organising committee, leaving the future of SACLA uncertain. Having realised the cancellation of the 2013 conference, Andre Calitz, Brenda Scholtz, and Clayton Burger from NMU decided to host the 2014 conference. Arrangements were made to obtain the seed money from the Botswana University, and steps were taken to start putting more formal structures in place.

Twenty-one full papers and four short papers were presented at the 2014 conference, held in the Summestrand Hotel, Port Elizabeth. Professor Steve Burges from the NMU Business School was the keynote speaker. Ten sponsors, including book publishers, supported the conference. The conference also awarded the first Best Paper Award. Professor

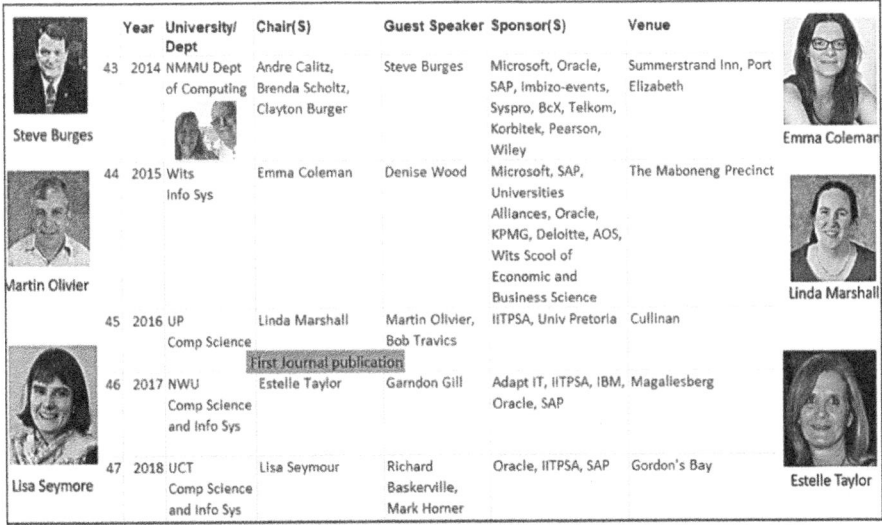

Fig. 12. SACLA hosting 2013 to 2018.

Janet Wesson from the Department of Computing Sciences at NMMU announced the SACLA 2014 Best Paper. The SACLA 2014 Best Paper Award went to Romeo Botes and Imelda Smit of North-West University for their paper titled "Renewal of a typical database systems module with NoSQL data stores".

The first HOD Colloquium was organised, bringing HODs together for discussions regarding the running of departments, staffing and common challenges experienced. Eleven HODs attended the colloquium with the assistance of 3 facilitators (Fig. 13).

The WITS Department of Information Systems hosted the 2015 conference, and Dr Emma Coleman was the conference chair. Guillaume Nel and Liezel Nel won the CS best paper award with their paper titled "Improving Program Quality: The Role of Process Measurement Data", and Thabang Serumola and Lisa F Seymour won the IS best paper award with their paper entitled "Factors Affecting Students Changing their Major to Information Systems". Dr Linda Marshall was the 2016 conference chair, and the first papers were selected for the journal publication that year. Professor Stefan Gruner from Pretoria assisted with the first Springer journal publication.

Professor Estelle Taylor from NWU chaired the 2017 conference, and Professor Martin Olivier (UP) was the guest speaker at the conference, speaking about computer forensics. Professor Lisa Seymour from UCT chaired the 2018 conference, held in Gordon's Bay. Seventy-seven papers were submitted, and 47 papers were accepted. Douglas Parry and Daniel Le Roux from Stellenbosch University received the conference's Best Paper Award for their "Off-task Media Use in Lectures: Towards a Theory of Determinants".

The 2019 conference took place at the Alpine Heath Resort in the Drakensberg, South Africa. Professor Mac van der Merwe was the conference chair [21]. The conference proceedings were compiled by Professor Bobby Tait and Professor Jan Kroeze. Three workshops were offered before the main conference, Academic Writing for Junior Informaticians and Computer Scientists, a South African Computer Accreditation Board

Back: Prof Jan Kroeze (UNISA), Prof Elize Ehlers (UJ), Prof Greg Foster (Rhodes)
Middle: MJ Matjuda (UL), Antoinette Lombard (VUT); Prof Bennet Alexander (UWC), Manoj Lall (TUT)
Front: Ntosh Wayi (UFH), Prof Lisa Seymour (UCT IS), Prof Jean Greyling (NMU), Prof Andre Calitz (NMU), Prof Jane Nash (Rhodes)

Fig. 13. First HOD Colloquium held in 2014

meeting and the Amazon Web Services- Educate and Academy Programs. Susan Campher, the previous SACLA Treasurer, facilitated registering SACLA as a non-profit organisation, which was why a Money Market account was opened as it was a requirement. SACLA is now a registered non-profit organisation with Registration No: 231-955 NPO.

SACLA 2020 was the first SACLA virtual conference, chaired by Professor Karen Bradshaw and Dr Ingrid Siebörger from RU. The 2020 conference had initially been planned from the 6th to 8 July at the Mpekweni Beach Resort in the Eastern Cape. However, in April 2020, the organisers realised that it would not be possible to hold an in-person conference due to the COVID-19 pandemic starting with arrangements for a virtual conference (Fig. 15). The conference organisers presented various awards at the end of the virtual conference, "Person who participated the most", "Delegate who asked the most questions" and Kudzai Katsidzira and Professor Lisa Seymour (UCT) won the Best Paper Award, with their paper entitled "Factors Impacting Using the Internet for Learning: The Digital Divide in South African Secondary Schools" [3] (Fig. 14).

SACLA celebrated 50 years of existence in 2021. The Academy of Computer Science hosted the SACLA 50th virtual conference and Software Engineering from UJ. The conference co-chairs were Professor Marijke Coetzee and Professor Wai Sze (Grace) Leung from UJ. The conference theme was *Post Pandemic Pedagogy* and focused on paper submissions that provided practical experiences and successes in computing education at a tertiary level to help CS, IS, and IT academics overcome challenges faced during the

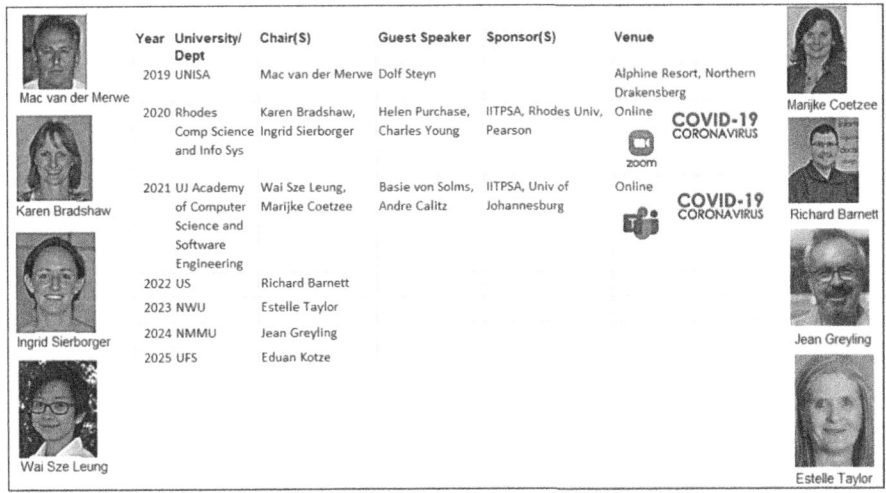

Fig. 14. SACLA hosting 2019 to 2025.

COVID-19 pandemic. Professor Basie von Solms presented the Keynote address entitled 'Computer Science – from then till now and into the future [22]. The conference was sponsored by the Institute for Information Technology Professionals from South Africa (IITPSA). In the IITPSA address at the conference, a call was made that the IITPSA consider a membership category for academics, called Professional Academic Member.

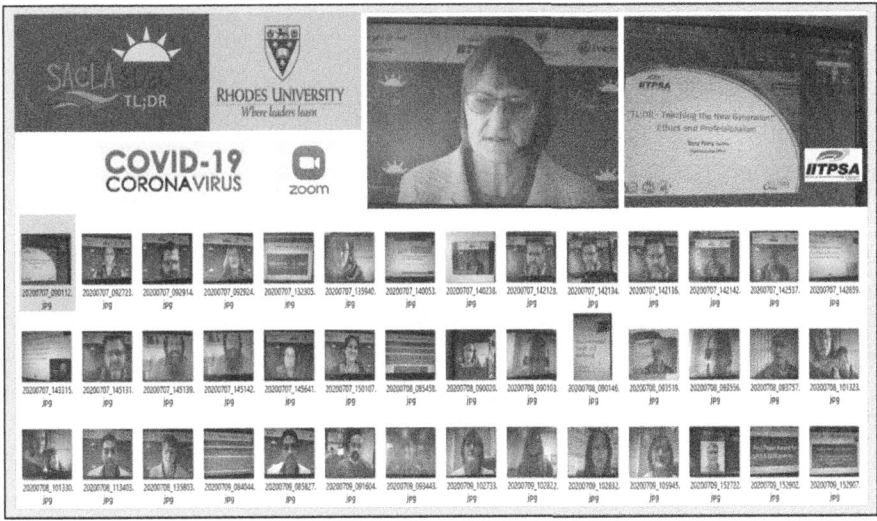

Fig. 15. SACLA 2020 virtual conference.

6 SACLA Paper Review Process, Publications and Conference Themes

The first call for papers for an annual SACLA conference usually takes place eight to nine months before the conference in the winter recess period in July each year. Full research papers and short papers are invited and are submitted using a double-blind peer-review process. Full papers and short papers, and position papers accepted after review are presented at the conference. SACLA conference proceedings with an IBSN number have been produced after the conference. At the 2009 conference AGM, delegates discussed the possibility of publishing selected papers in a journal and decided that the editor of SACJ, Professor Lukas Venter, would be approached. Since 2016, the SACLA conference organisers have produced two publications annually, namely:

- The SACLA Conference proceedings: Selected papers that are not included in the journal publication are published in the conference proceedings; and
- A Springer Journal publication: Communications in Computer and Information Science (CCIS Vol. 730) were selected top papers (up to 20%) presented at SACLA are published in the journal.

The themes of conferences are presented in Table 1, and the number of papers accepted and a number of attendees is presented in Table 2.

Table 1. SACLA conference themes

Year	University	Themes
2009	Rhodes	Connecting Students to the Future
2010	UP	Scoring IT education goals in 2010
2012	UFS	Educating in a changing environment
2014	NMU	ICT Education in the Cyber World
2015	WITS IS	Renewing ICT teaching and learning: Building on the past to create new energies
2016	UP CS	Achieving Brilliance in ICT Education
2017	NWU	Keeping Education Relevant: Infinite possibilities
2019	UNISA	Computing Matters of Course!
2020	Rhodes	TL; DR: Teaching the New Generation!
2021	UJ	Post-Pandemic Pedagogy

Table 2. Publications

Year	Total received	Accepted	Springer journal	Proceedings	Short	Attendees
2012	23	19 (83%)		9 (47%)	10	33 (AGM)
2014	30	25 (83%)		21 (84%)	4	56

(*continued*)

Table 2. (*continued*)

Year	Total received	Accepted	Springer journal	Proceedings	Short	Attendees
2017	63	40 (63%)	22 (35%)	18 (29%)	0	39 (AGM)
2018	77	47 (61%)	23 (30%)	24 (31%)	0	35 (AGM)
2019	59	27 (46%)	16 (27%)	11 (19%)	4	58
2020	53	20 (38%)	13 (24%)	8 (15%)	6	110 virtual
2021	23	10 (43%)	10 (43%)		0	72 virtual

7 SAICSIT and SACLA

The South African Institute for Computer Scientists and Information Technologists (SAICSIT) was established in 1982 [15]. The main aims of SAICSIT (2021) are to:

- Enhance research and development collaboration with other professional bodies;
- Support education and training;
- Seek solutions to technical and socio-economic problems in South Africa by means of research and development projects; and
- Strive for professionalism and excellence.

A primary strategic activity of SAICSIT has been the annual conference, which started in 1981 (predating the formation of the Institute). SAICSIT founded and managed the South African Computer Journal (SACJ). SACJ is an accredited specialist IT academic journal, publishing research articles, technical reports and communications in the Computer Science and Information Systems domains. A more significant number of academics belong both to SAICSIT and SACLA. The SACLA President is a member of the SAICSIT council. Both organisations request academics to support both conferences annually.

At the 2012 SACLA AGM, a proposal was discussed to host the SAICSIT and SACLA conferences at the same venue, one after the other, not in parallel. Delegates agreed that each conference must retain its own identity. The AGM minutes indicate, "We do not want to combine the conferences because the purposes are different".

8 The Social Side of SACLA

The SACLA conferences included welcome functions and gala dinners. The gala dinners included the Best Paper Award and an auction, started in 2014, by Professor Jean Greyling from the Nelson Mandela University, called "Conference with a conscious". The 'Conference with a conscious' idea was to identify a charity, obtain some sponsored and purchased artefacts, and have the delegates bid for an artefact. The proceeds would then be donated to the charity.

The late Professor Derick Smith was the guest speaker at SACLA 2009, and the delegates included Rhodes Chris Upfold, John McNeill and Karen Bagshaw (Fig. 16).

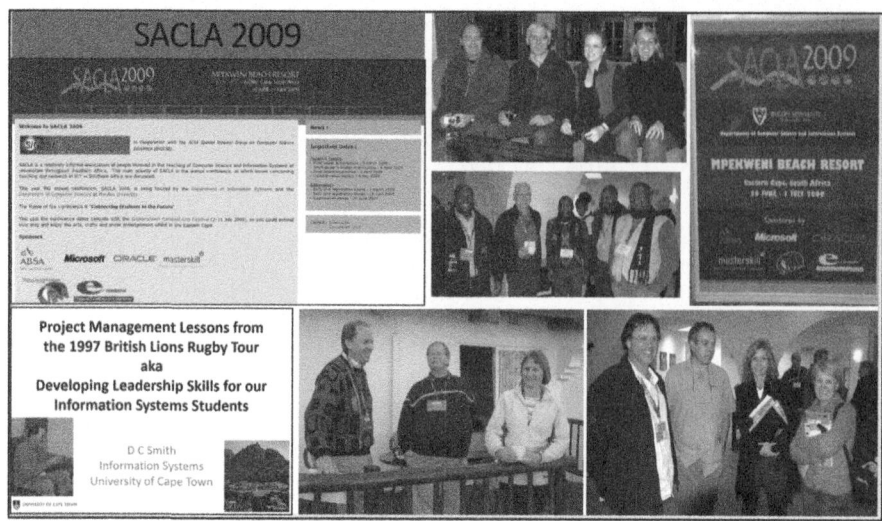

Fig. 16. SACLA 2009 attendees

In the top middle photo is Professor Jan Kroeze (UNISA), the late Pieter Joubert (Sr, UP), Sonja Visagie (neé Cloete, UP) and Riana Steyn (UP). Professor Mike Hart attended with UCT delegates. Delegates from NMU included Professors Jean Greyling, Brenda Scholtz and Charmain Cilliers.

SACLA 2019 was held in the Alpine Heath Resort, Drakensberg, Natal. The Best Paper Award was presented to Pakiso Khomokhoana and Liezel Nel from NWU (Fig. 17). The gala evening was well supported, with delegates supporting the *conference with a conscious* fundraiser and later group dancing.

Fig. 17. SACLA 2019 conference

9 Future of SACLA

The Covid-19 pandemic has changed the format of the past two SACLA conferences, the SACLA 2020 and SACLA 2021 conferences were both virtual conferences. Future SACLA conferences will have to cater for face-to-face presentations and possibly include virtual presentations. The pandemic has introduced a new era of on-line and hybrid teaching approaches, examining and research supervision challenges, which will shape the future trajectory of SACLA conference themes. The SACLA 2022 conference is being hosted by the University of Stellenbosch in July 2022 in Cape Town, followed by North-West University in 2023.

References

1. Anderson, B.: SACLA history (2021)
2. Bishop, J.: First years of SACLA. Department of Computer Science, University of Stekllenbosch (2021)
3. Bradshaw, K.: Computer Science Department, Rhodes University, South Africa (2021)
4. Calitz, A.P., Cowley, L., Petratos, S.: A proposed structure for managing IT diplomas' programme content in South Africa. In: SACLA 2020 Conference, South Africa, 6–8 July 2020 (2020)
5. Coetzee, M.: RAU Rapport No. 2, 3 September 1969 (2021)
6. Fischer, B.: Science faculty centenary yearbook (2021)
7. Kotze, E.: University of Free State Computer Science (2021)
8. Kourie, D.: University of Pretoria Computer Science (2021)
9. Kroeze, J.H.: School of Computing History - UNISA (2021)
10. Marshall, L.: University Pretoria Computer Science Department (2021)
11. Newcastle University: Computing Science History Seminars (2021). https://www.ncl.ac.uk/computing/about/history/seminars/
12. Olivier, M.: Department Computer Science University Pretoria (2021)
13. Rautenbach, T.: History of the Department of Computing Sciences at the Nelson Mandela University (2017). Unpublished publication
14. SACLA: Southern African Computer Lecturer's Association (2021). http://www.sacla.org.za/
15. SAICSIT: The South African Institute for Computer Scientists and Information Technologists (2021). www.SAICSIT.org
16. Shallit, J.: A Very Brief History of Computer Science (1995). https://cs.uwaterloo.ca/~shallit/Courses/134/history.html
17. Suleman, H.: Computer Science at UCT: 1970–1990 (2021)
18. Taylor, E.: PU vir CHO – Rekenaarwetenskap (2021)
19. Van den Heever, R.: Workgroup for Computer Education Report of a Conference held at the Golden Gate National Park, 29–30 November 1973 (2021)
20. Wells, G.: Rhodes University Department of Computer Science (2021)
21. Van der Merwe, M.: Department of Information Systems, UNISA, South Africa (2020)
22. Von Solms, B.: Computer science – from then till now and into the future. In: SACLA 2021 Conference, 15–16 July 2021 (2021)
23. Von Solms, R.: TECLA: Teams meeting 2 September 2021 (2021)

Teaching Innovation

Minimising Tertiary Inter-group Connectedness Over Successive Rounds

Andrew Broekman(✉) [iD] and Linda Marshall[iD]

Department of Computer Science, University of Pretoria, Pretoria, South Africa
andrew.broekman@up.ac.za, lmarshall@cs.up.ac.za

Abstract. Rocking the boat is a teaching strategy to rapidly teach tertiary computer science students the required group and communication skills for software engineering. This strategy proposes the introduction of high-risk factors into the group dynamics over short time periods. Group instability is regarded as a risk factor. It is introduced by reshuffling groups between successive rounds. The main examples of allocation methods applied during reshuffling include random allocation, academic standing, participation level and Belbin roles. The reshuffling of groups should ensure that subsequent groups remain heterogeneous with regards to contact between students, that is minimise the inter-group connectedness. Current group formation methods and related software do not focus on the inter-group connectedness over successive group allocations. The construction and tracing of groups by hand to ensure a minimum inter-group connectedness is time-consuming and prone to error. This paper provides a genetic algorithm from the subset of evolutionary algorithms to minimise inter-group connectedness. The proposed algorithm reduces the time and error in constructing groups based on random allocation over successive rounds.

Keywords: Group formation · Teaching teamwork · Evolutionary algorithms · Genetic algorithm · Optimisation

1 Introduction

Enrolled B.Sc. Computer Science (CS) students at the University of Pretoria are required to complete a Software Engineering (SE) module in their final year. The mentioned module aims to equip students with the required skills to execute the development of a software system based on SE principles. Each year industry is invited to submit project proposals to be implemented by students as a capstone project, also known as Project-Based Learning (PBL). The capstone project requires a group[1] of five students to iterate through the Software Development Life Cycle (SDLC) with their choices of software development teaching strategy, such as agile, scrum, or eXtreme Programming (XP). Students are expected to proffer a working software system with all associated software artefacts at the end of the academic year. A non-exhaustive list of artefacts includes Unified Object Modelling (UML) diagrams, Software Requirement Specifications (SRS),

---

[1] For this paper, the word *team* can be used interchangeably for *group*.

Software Configuration Items (SCI), Project Management (PM) with the necessary Continuous Integration (CI) and Continuous Deployment (CD) on a version-controlled repository.

Working together in the same group for a year presents students with various social, behavioural, and emotional challenges. The *Rocking The Boat* (RTB) teaching strategy provides students with the requisite skill set. The strategy requires the introduction of high-risk factors into group dynamics. Skills are taught by necessitating students to confront and resolve conflict caused by the introduced risk factors. In the aforementioned SE module, this is accomplished by requiring students to abruptly iterate through the SDLC and implement a software system under high stress with a low-stake approach in terms of their assessment. This approach is presented to students through a *mini-project* [11, 16–18].

The mini-project is a medium to large scope project prepared by the lecturing staff and presented during the first six weeks of the academic year. Students are expected to iterate through the SDLC using the waterfall software development methodology.

RTB introduces various risk factors, one of which is *instability* [16, 18]. A change in the membership structure of a group causes this instability. For multiple rounds (allocations) in an evaluation period, the contact between group members should be minimal, thus keeping the long-term effect of negative group dynamics at a minimum. Peer review and self-reflection after a round enable students to vent any negative emotions and learn from the experience. Minimal contact between members ensures high instability in the groups. This paper introduces an evolutionary algorithm to minimise the contact between individuals over an evaluation period. For this paper, the focus is on automating group allocations and not on the specific evolutionary algorithm. The presented approach focuses on random group allocation using a Genetic Algorithm. It can be extended to automate other allocation methods. Automating group formation in the context of RTB decreases administrative overhead and error.

Section 2 presents the reader with an overview of RTB teaching strategy. Section 3 highlights the effort needed to manually construct groups for the application of this strategy. An overview of Genetic Algorithms (GA) is provided in Sect. 4. Section 5 presents the implementation with an outline of future work in Sect. 6, and concluding remarks are presented in Sect. 7.

2 Rocking the Boat (RTB)

In 2011 Pieterse et al. [16] introduced RTB, after acquiring ethical clearance, as a teaching strategy to teach tertiary CS students the required group and communication skills for SE. The strategy introduces high-risk factors over a short period into the group. A short period is specifically selected to ensure that there are no lasting effects. Students learn the required skills practically by needing to solve group conflicts to accomplish a stated academic goal [16, 18].

2.1 Theory

To successfully teach group dynamics, the group needs to experience four stages: (a) *Forming*—Initial boundaries are determined by familiarising oneself with the group; (b)

Storming—Group hierarchy and division of roles are established through conflict; (c) *Norming*—Rules of engagement and group cohesion are established through the resolution of conflict; and (d) *Performing*—Functioning as a single unit to accomplish a common goal. Iteration through the first three stages is a prerequisite to reaching the final *performing stage* [16, 22]. RTB aims to introduce multiple storming and norming stages to accelerate the process of learning the skill set required for performing in a group [16, 18]. Students learn group management, communication and conflict resolution skills at a rapid pace because of the introduced risk factors.

RTB risk factors include: (a) *Large group*—Small groups should not exceed five members; (b) *Instability*—Frequent changes to the group membership structure; (c) *Large project scope*—Requiring the completion of a complex goal in minimal time; (d) Academic diversity—Ensuring the group is composed of students with varying academic skills; and (e) *Lack of experience*—Execution of a goal in a context where the group has little or no experience [11, 16, 18].

2.2 Participation Levels

The level of engagement of a member within a group is referred to as the *participation level*. The *Ringelmann effect* refers to the negative correlation between group size and the level of engagement [7]. RTB identifies four types of participation levels: (a) *Diligent isolate*—"An individualistic member that relies only on himself/herself to complete the tasks at hand"; (b) *Insightful shaper*—"A member who takes great responsibility to ensure that the required work gets done and usually works harder than is expected"; (c) *Compliant worker*—"An individual who is likely to accept the decision of others without consideration"; and (d) *Social loafer*—"An individual whose contribution is perceived to be inferior to that of others in a team" [11, 16, 18].

2.3 Group Allocations

Section 2.1 provided an overview of various risk factors. Instability is introduced by changing the group members in a group [8, 18]. Two parties that have worked together in a group is said to have had *contact*. Contact between two members should ideally only occur at most once. Placing members together who had prior contact reduces the instability of the group [11].

RTB presents various group allocation strategies, namely random allocation, participation levels, academic standing and Belbin roles, discussed below. There is no requirement that the same allocation method is used across different group formations over the rounds.

Random Allocation. Members are randomly allocated to groups. Random allocation ensures that an equal probability exists for the group to be functional or dysfunctional. It is explicitly stated that other factors might contribute to dysfunction in the group. Only factors that the lecturing staff can control is taken into consideration [16].

Participation Levels. Group members are allocated based on participation level as measured by some instrument, possibly peer review on a previous round. Students can either be grouped on the same participation level; that is to say, all diligent isolates are placed together in groups, followed by social loafers, insightful shapers and lastly, compliant workers [16]. An alternative allocation strategy is to ensure inter-group balance where the group is composed of a balanced allocation of participation levels [15].

Academic Standing. *Academic diverse groups* are constructed by ranking the marks of the cohort from highest to lowest. How academic marks are combined per student is outside the scope of this paper. Upon completion of ranking the students, the top n students are allocated to n different groups. This process is repeated until all students have been allocated to groups. After group formation, the average and standard deviation for each group are calculated using the marks of the individual group members. Members may be moved between groups to ensure a similar average and standard deviation across groups.

Alternatively, *academically similar groups* can be constructed. After students have been ranked based on academic marks, the top n students are allocated to the first group, whereafter the process is repeated until all students are divided into groups [11, 15, 18].

Belbin Roles. Belbin's theory focuses on the behaviour of individuals when present in a group and the role that they fulfil with the group hierarchy. Groups can be constructed in such a way as to be Belbin balanced, implying that the group organisation satisfies Belbin's recommendations. Alternatively, groups can be constructed to be Belbin unbalanced [11].

Hybrid Allocations. Hybrid allocations make use of a combination of the above allocation strategies. An example strategy is to classify students using two orthogonal classifiers. The first axis is based on participation and is divided into three quadrants corresponding to leader, worker or slacker. The second orthogonal axis is based on academic ability and is again divided into three quadrants corresponding to strong, average, and weak [16, 18].

2.4 Application

The mini-project is presented using an adaptation of the waterfall approach due to time constraints. The project proposal is prepared by the lecturing staff, who assumes the role of the client during the lifetime of the mini-project.

A presentation is presented to students that describe various aspects of the project. Aspects that are covered include, but is not limited to, the vision, mission, purpose, target audience, platforms and proposed technologies. Students are requested to elicit requirements from the client after the presentation.

The staff proceeded to allocate students into groups larger than five students using one of the allocation strategies. This large group size is the first risk factor being introduced. The first round requires students to submit a Software Requirements Specification (SRS) document created in LaTeX. The document should only provide information on the functional requirements of the proposed software system. The submitted SRS document should contain the following concepts: (a) *Introduction*—How the group understands the system; (b) *Functional requirements*—Functional requirements of the system including use-cases; (c) *User characteristics*—Users who will engage with the system; (d) *Traceability matrix*—Ensuring coverage of the requirements by the identified use-cases; and (e) *Class diagram*—Proposed syntactically correct UML class diagram. The large scope of work is the second risk factor being introduced.

Second group allocations and instructions for the round are released twenty-four hours before the first round deadline. Groups are reshuffled for the second round. The reshuffling of groups introduces the third risk factor of instability. Groups are expected to create a new SRS document from their collective first submissions. The newly created SRS document is to include the non-functional requirements of the system, namely: (a) *Quality requirements*—Discuss and quantify the non-functional quality requirements required by the system; (b) *Architecture patterns and constraints*—Discuss the choice of patterns; (c) *Actor-system interaction models*; (d) *Deployment model*—Discussion in addition to a syntactically correct UML deployment diagram; and (e) *Technology choice*—Discuss mapping of the technology-neutral solution to a technology stack.

Upon completion of the second round, the lecturing staff release the model solution SRS document. During the subsequent lecture, students are provided with the differences between software development and SE. A high-level outline of enterprise software architectures and frameworks is also provided. After the presentation, students are afforded the opportunity to ask the lecturing staff about the provided model solution.

For the final phase students are divided into a two-tier grouping, depending on headcount. The first tier represents the virtual organisation for which the students will work. The second tier represents the subsystems identified by the lecturing staff. An integration group is constructed out of students who achieved the highest marks in the prerequisite modules, namely *Software Modelling* and *Data Structures and Algorithms*. The integration group oversees the virtual organisation. The integration group has two functions namely: (a) Facilitate communication across subsystem groups; and (b) Required to implement the web and mobile application. The web and mobile application is built in a hybrid technology such as Ionic Angular.

2.5 2021 Group Allocation Requirements

In the 2021 academic year, the students received the following brief: *CrowdBook is a crowd-sourcing book application that will allow people to search for books, be alerted when selected books are placed on the platform enabling them to rent or buy books. The current market of online booksellers is catering to individuals who only want to purchase new books or are too scared to buy second-hand books because it is difficult to establish the book's quality remotely. This application hopes to target book lovers from all walks of life who still loves the feel and smell of a book on a lazy Sunday afternoon relaxing in a hammock.*

During the requirement elicitation, the client stated that the platform would use the Bitcoin Testnet [21] to facilitate payment between parties. More specifically, the software system was functioning as an escrow account, and that the blockchain should be monitored directly by the software in question. That is no external Application Programming Interface (API) should be used to interrogate the blockchain.

For the Crowdbook system the following subsystems were identified and presented to the students: (a) Book subsystem; (b) Notification subsystem; (c) Payment subsystem; (d) Recommendation subsystem; (e) Reporting subsystem; (f) Shipping subsystem; and (g) User subsystem.

Table 1. Student numbers for SE module.

Year	Student numbers
2011	64
2012	59
2013	54
2014	56
2015	98
2016	75
2017	96
2018	81
2019	79
2020	123
2021	153

Table 2. Table illustrating the various group allocation parameters over the academic years of 2020 and 2021.

Round	Activity	Year	
		2020	2021
Round 1	Number of groups	22	19
	Group size	6	8
Round 2	Number of groups	17	19
	Group size	8	8
Round 3	Number of major groups	3	3
	Integration group size	6	6
	Number of subsystems	6	7
	Subsystem group size	7	6
	Total groups	21	24

The lecturing staff required group allocations for three rounds. The first and second-round groups were to be of size eight and allocated using the random group allocation method. For round three, a two-tier division is required due to the class size. The class is divided into three development groups, namely Alpha, Beta and Gamma, referred to as the first tier. For the second tier, each of the first-tier development groups is further divided into eight groups—seven subsystems and an integration group. Each of these groups comprises six members. The integration group should be constructed using the balanced academic standing group allocation method. The remaining group are allocated using the random group allocation method. Table 1, read in conjunction with Table 2, highlights the complexity of constructing and tracing groups by hand.

3 Allocating Groups

Various software systems [2, 3, 24] exist, and papers have been published on group allocation algorithms. Examples of such algorithms include ant colony optimisation [4], evolutionary algorithms [6, 12, 13, 25], particle swarm optimisation [10, 26] and those using more traditional mathematical constructs [2]. Limited literature on automating group allocation such that the inter-group connectedness over successive rounds is minimised exists.

3.1 Automation

For the presented SE module at the University of Pretoria the allocation of students is done manually. This allocation involves the manual tracing of students, either using pen and paper or a spreadsheet with complex formulae taking many hours, to ensure that contact is minimised over rounds. While a manual approach can be used for smaller groups, the SE module has seen an increase in student numbers from 2020 (refer to Table 1)[2]. This increase in student numbers presents the lecturing staff with a scaling problem.

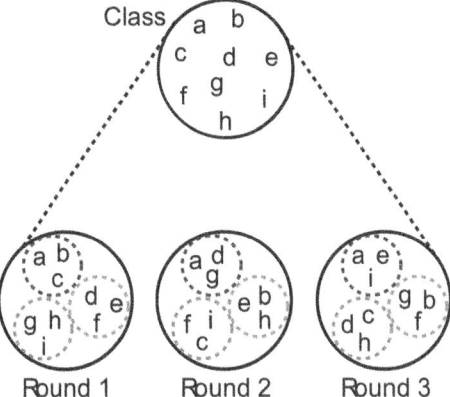

Fig. 1. Visual example showing group allocations over successive rounds with an inter-connectedness value of one.

3.2 Inter-group Connectedness

Suppose a class of nine students, *indicated by the letters a – i,* in Fig. 1, need to be allocated to three rounds with each group comprising of three students. The first-round students are divided into groups as indicated by the container named Round 1. Coloured circles indicate the respective group allocations. For the second round, students need to be divided, with the constraint that if two students have worked together in Round 1, they may not work together again in Round 2. Similarly, for Round 3, students need to be allocated again such that students have not previously worked together in Round 1 or Round 2.

---

[2] Student numbers for the years 2011–2013 (inclusive) is obtained from literature [11]. Figures from 2014–2020 (inclusive) refer to the student numbers at the end of the module. The figure for 2021 is as at the fourth round of the mini-project.

When two students work together in a Round, say students d & e in Round 1, then such students are said to be *connected*. If overall rounds, any two students only had contact once, then such a group allocation is said to have an inter-connectedness value of one, the minimum value for any group allocation. This paper seeks to minimise inter-connectedness, ideally to a value of one.

Formally to partition a class C into a set of rounds $R = \{R_1, R_2, R_3, \ldots, R_n\}$ where $n \in \mathbb{N}, n > 0$. Each round R_i distinctly partitions the class C such that $R_{i1} \cap R_{i2} \cap R_{i3} \cap \ldots \cap R_{im} = \emptyset, 1 < i < n, \forall m$, where $i, m \in \mathbb{N}$ and $m > 0$. Furthermore, each round, R_i partitions the class C uniquely such that once two elements $x, y \in C$ co-occur in a group $R_{ij}, j \in \mathbb{N}, j > 0$, they never co-occur in another group $R_{kp}, \forall k, p$ with $i \neq k$.

4 Genetic Algorithms

GAs is a subset of Evolutionary Algorithms (EA), which are metaheuristic optimisation algorithms inspired by biological evolution. GAs are based on the theory of evolution—Survival of the fittest—from Charles Darwin. GAs operates on a population of individuals, where each individual represents a possible solution. Each subcontract will be discussed to promote understanding in the subsequent sections. Various biological subconstructs are mimicked during each iteration of the algorithm, such as reproduction, mutation, crossover, and selection, each discussed in greater detail in subsequent sections. To drive the algorithm towards a candidate solution, a fitness function is employed [1, 20].

4.1 Representation

The representation of the solution plays a critical role in the design and functioning of the genetic algorithm. Representation is concerned with the encoding of the arguments which constitute the solution parameters. A representation is problem-dependent. The design of a good representation takes the constructs of crossover and mutation into account. Various representation formats are commonly used which include, arrays, bit representations, strings [20] and even more complex structures like graphs [5, 14] or Artificial Neural Networks (ANN) [9, 23].

4.2 Crossover and Mutation Operators

The canonical definition of a GA defines crossover operators as global operators. The crossover operators are used to ensure that the solution space is thoroughly explored. The crossover operators are concerned with how the representation of two or more selected individuals can be combined to produce new offspring. It is this recombination that provides the algorithm with the ability to *explore* the solution space [20].

Mutation operators *exploit* a given area of the fitness landscape to determine the most optimal solution by operating on the constituent parts of the individual [20].

4.3 Selection

Two stages in the canonical GA require the selection of individuals. The first stage occurs when several individuals need to be selected for the crossover operator [20]. It is mentioned explicitly that the number of individuals selected must be equal to the number of individuals required by the operator. The second stage of selection occurs when a new population must be created for the next epoch (generation) of the algorithm [20]. Some implementations of GAs do not select a new population, but rather cull the entire current population and replace it with the newly created offspring. The argument against this approach is that good genetic material may be lost in the culling process as offspring are not guaranteed to be fitter than their parents because of the stochastic nature of the algorithm. The alternative is to combine the current population and offspring into a new cohort and apply a selection method to the newly combined cohort.

The selection algorithm is one of the mechanisms to control the convergence rate in the GA. The convergence rate refers to the rate that the GA transitions from exploration of the solution space to the exploitation of the solution space to finally produce a candidate solution [19, 20]. When searching for solutions by using artificial intelligence or machine learning, there is an interplay where the solution space should be explored at the start of the algorithm. As the algorithm progresses towards the end, the algorithm should ideally have found a "good" solution, possibly the global optimum, and start to exploit that region to obtain the best candidate solution.

Selection policies can be ordered on a spectrum by the speed at which the GA converges. *Random selection policy* is the slowest selection policy in which an individual is selected randomly from the population without any regard to fitness [20]. The effect of this approach is that the algorithm will either converge extremely slowly or never. The lack of convergence is because the best genetic material is not carried forward to the subsequent populations. On the other end of the spectrum, with the *elitism selection policy*, only the fittest individuals are selected [20]. This causes the GA to converge extremely quick in most cases, called premature convergence. Premature convergence causes the algorithm not to explore the solution space sufficiently often, causing one not to find a "good" candidate solution, according to the fitness function.

The selection policy *tournament selection* is located between the two previously mentioned selection policies. Tournament selection proceeds to select a subgroup from the cohort of size N, where N is called the arity of the tournament selection. Once the subgroup has been selected, the fittest individual is then selected from this group [20]. The arity controls the convergence of the algorithm, also known as the convergence pressure. If the arity is equal to one, tournament selection becomes random selection, whereas if the arity is equal to the cohort, it becomes the elitist selection.

4.4 Termination

GAs have no idea as to whether a global optimum solution has been reached. A mechanism is required to determine when the algorithm should be terminated. One approach is to observe whether evolution has taken place for the past N generations. The value of N would be problem dependent. If no evolution has occurred, it can be concluded that the algorithm has obtained a candidate solution. An alternative approach might limit the algorithm to a prespecified number of generations or to a predetermined execution time. The disadvantage of this approach is that one does not necessarily know the required number of generations or time for a candidate solution to be found beforehand. An alternative is to terminate the algorithm when a fit enough individual has been found. For some problems, it might be easy to determine what this value should be, while for other problems, it is unknown beforehand what the threshold value should be [20].

4.5 Fitness Function

The fitness function is critical to the success of the GA as it guides the algorithm in finding a candidate solution [20]. The fitness function might be easily determined and computed for some problems, such as a set of simultaneous equations. For other problems, it might be intractable or computationally expensive to calculate the fitness, such as wind tunnel simulations for motorbike frames or building plans for an office skyscraper.

5 Application of a GA for Inter-group Connectedness

The approach taken to implement the 2021 group allocation requirements (Sect. 2.5) using a GA is given using an example with mock data to demonstrate the algorithm.

5.1 Data Overview

A Comma Separated Value (CSV) file is used to feed the required input data into the GA. The file is obtained from the University of Pretoria's student database record system and is combined with the marks for prerequisite modules. Data points for the file include: (a) *Emplid*—The students' employee identifier on the human resources system. Functions as a "student number"; (b) *Last Name*—The family name of the student as captured by the human resources system; (c) *Initials (Name)*—The given name including the initials of the student as captured by the human resources system; (d) *Title*—The title of the student as captured by the human resources system; (e) *Email Address*—The email address of the students as submitted in enrollment for the year by the student; (f) *COS212*—Prerequisite *Data structures and Algorithms* module; (g) *COS214*—Prerequisite *Software Modelling* module; and (h) *Total*—An external weighted column which was skewed 60% towards the Data structures and Algorithms module with the Software Modelling module contributing 40% towards the weighted average.

5.2 Methodology

The strongest students of the group, based on previous year marks, to be represented in the various "organisations" integration groups is required. For this reason, the problem is approached by first allocating the last round and working towards allocating the groups for the second and then the first round. The allocation for the third round is done programmatically by ranking students based on the weighted marks from high to low. The top three students are selected and allocated to the Alpha, Beta, and Gamma integration groups respectively. Thereafter the following three students are allocated to the respective integration groups. This process is continued until each integration group is composed out of six students. The remaining students are randomly allocated to the respective seven subsystem groups. After allocation, three students remained, which were equally distributed between the larger first-tier groups. Each of the remaining students was however randomly allocated to one of the seven subsystems. The reason for allocating the last round in this programmatic way was to minimise the solution space that the GA had to search.

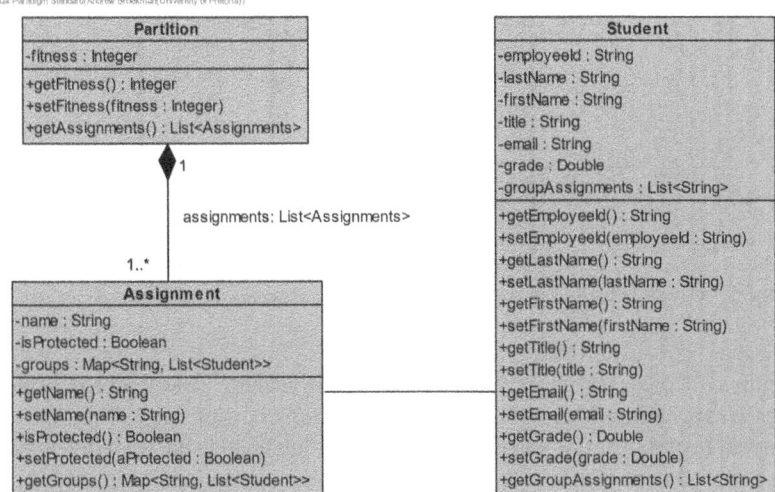

Fig. 2. UML class diagram illustrating the design used to represent the individuals in the GA

For the GA an individual is represented by containing a list of three assignments which contains a Boolean flag, representing whether the groups in the assignment can be mutated. Each assignment has a map of groups, which maps the group to a list of students. For a visual reader, refer to the attached class diagram in Fig. 2. The *Partition* class represents the individual in the algorithm. For each round a corresponding *Assignment* class is created together with a *Student* class to represent students.

Consider two randomly shuffled class lists, C_1 and C_2, upon which a crossover operator is applied. A single point crossover operator will select a random index. This random index will function as the pivot point around which the two lists will be recombined. Class lists C_1 and C_2 will be split to obtain four parts namely $C1 = \{C_{11}, C_{12}\}$ and $C_2 = \{C_{21}, C_{22}\}$. After which the crossover operator will produce two new class lists, namely $C_3 = \{C_{21}, C_{12}\}$ and $C_4 = \{C_{11}, C_{22}\}$.

Since the lists C_1 and C_2 are randomly shuffled, the newly constructed lists, C_3 and C_4, will be corrupt, that is students will be duplicated while others are missing.

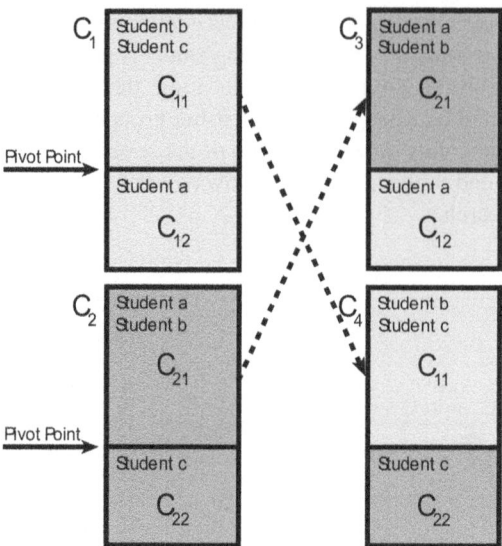

Fig. 3. Visual depiction of one point crossover operator corrupting student class lists

Figure 3 depicts the above-explained scenario visually. The figure indicates how *Student a* is appearing twice in class list C_3 while *Student c* is missing. A similar scenario is observed in class list C_4, where the *Student b* is duplicated and *Student a* is missing.

To correct an error in an individual would involve comparing the students in the list against some reference list. This would involve a multi-step process: (a) determine the set difference between each assignment and the reference list of students; (b) determine which students are duplicate; (c) removing duplicate students; (d) determine the set difference again between each assignment and the reference list; and (e) adding the calculated set difference back into the list of students for each assignment. For this reason, all crossover operations are removed from the algorithm; in other words, only mutation operators were utilised.

Two mutation operators were defined, namely (a) Removes a random student from the selected group and place them into a random target group; and (b) Removes a random student from a randomly selected target group and adds them into the currently selected group.

For this paper, a combination of three termination policies was used: (a) no evolution; (b) generation limit; and (c) fitness threshold value.

5.3 Mock Data Example

A class list with 20 mock students was prepared. The parameters used during the run for this example are as follows: (a) *Number of assignments*: 3; (b) *Group size*—4; (c) *Population Size*—50; (d) *Mutation Probability*—0.10; and (e) *Tournament Arity*—50. The stated requirement for this example is to divide students randomly into groups of size four, such that the inter-group connectedness is minimised over three successive rounds. The group sets for each round is displayed in the following example:

Class = { 01, 02, 03, 04, 05, 06, 07, 08, 09, 10,
11, 12, 13, 14, 15, 16, 17, 18, 19, 20 }
Rounds = { Round 1, Round 2, Round 3 },
Round 1 = { Group 1, Group 2, Group 3, Group 4, Group 5 },
= { { 05, 09, 13, 20 }, { 02, 04, 15, 19 }, { 01, 08, 14, 17 },
{ 03, 10, 16, 18 }, { 06, 07, 11, 12 } },
Round 2 = { Group 1, Group 2, Group 3, Group 4, Group 5 },
= { { 01, 02, 11, 13 }, { 04, 07, 08, 16 }, { 09, 10, 17, 19 },
{ 03, 06, 15, 20 } }, { 5, 12, 14, 18 } }
Round 3 = { Group 1, Group 2, Group 3, Group 4, Group 5 },
= { { 01, 07, 18, 19 }, { 02, 05, 06, 10 }, { 09, 14, 15, 16 },
{ 04, 11, 17, 20 }, { 03, 08, 12, 13 } }

Table 3 provides a visual investigation into the connectedness between students. A 10% grayscale background indicates the groups for the first round, a 40% grayscale background indicates the second round, while a 70% grayscale background indicates the last round. The count indicated in the table indicates the number of times that there was

Table 3. Table displaying the inter-group connectedness visually over three successive rounds. The first round is indicated in 10% grayscale, the second in 40% grayscale and the third in 70% grayscale.

	01	02	03	04	05	06	07	08	09	10	11	12	13	14	15	16	17	18
01		1					1	1			1		1	1	1		1	1
02	1			1	1	1					1	1		1	1			
03						1		1		1		1	1		1	1		1
04		1					1	1			1				1	1	1	
05		1		1				1	1		1	1	1					1
06		1	1		1		1			1	1	1			1			
07	1		1		1		1				1	1			1			1
08	1		1	1			1				1	1	1			1	1	
09				1						1			1	1	1	1	1	
10		1	1		1	1			1						1	1	1	1
11	1	1		1		1	1					1	1				1	
12		1		1	1	1	1			1			1	1				1
13	1	1	1	1			1	1			1	1						
14	1			1			1	1			1				1	1	1	1
15		1	1	1	1			1						1		1		
16			1	1			1	1	1					1	1			1
17	1			1			1	1	1					1				
18	1		1		1				1			1		1		1		
19	1	1		1			1	1							1		1	1
20			1	1	1	1		1			1			1			1	

contact between the two students over different rounds. As can be noted from the table, all students had precisely only one interaction with each other. It is stated that the table is symmetrical around the diagonal, as is expected. If student 2 and 3 work together, then students 3 and 2 also work together. When plotting the connectedness between students, it is only necessary to plot the upper or lower quadrants. For this paper, both the upper and lower quadrants of the triangle were completed.

Either the row or column of a student in question can be used to lookup with which other students in other groups they have contact. To understand with whom student 08 has contact in Round 1, the row or column corresponding to student 08 is used for the lookup. Round 1 corresponds to all students coloured in 10% grayscale, thus {01, 14, 17}. All students coloured in 40% grayscale correspond to Round 2 which is {04, 07, 16}, while the group of students for Round 3 is provided by {03, 12, 13} with a 70% grayscale background.

6 Future Work

Due to corrupting the student lists crossover operators were removed. If a traditional crossover operator is used, a repair mechanism would need to be run after crossover to repair the integrity of student lists. Future work needs to investigate how crossover operators could be defined for the lists to reduce the convergence time by searching a more extensive solution space while not corrupting the integrity of the lists. If a crossover operator cannot be defined, how would a repair mechanism function? Would the introduced overhead of such a repair mechanism justify the additional solution space searched?

Further work is also to be undertaken in expanding the capability of the GA to allow for more complex optimisations such as using varying group allocation strategies over different rounds as being able to understand, interpret and allocate students according to various personality frameworks such as Belbin roles and Briggs Myers personality types. Moreno et al. [12] proposed a GA to group students on characteristics for a single round. This work can be considered for extension to allow for the grouping of students over different rounds with various characteristics while ensuring the inter-connectedness is minimised.

Migration patterns [11] will be investigated at the completion of the mini-project. Once groups for the capstone project have been formed, an investigation into the migration patterns will take place to compare the current academic year migration patterns to those presented in previous years. It is believed that these patterns could assist lecturers in identifying at-risk groups using automated software.

7 Conclusion

This paper presented the reader with a high-level overview of the Rocking The Boat teaching strategy and Genetic Algorithms. With the increase in student numbers experienced by the lecturing staff of the Software Engineering module, an alternative solution was required to construct groups that still satisfy the instability risk mandated by the strategy. This paper proposed using genetic algorithms to construct groups by using a random allocation strategy while ensuring that the inter-connectedness across successive rounds is minimised. This approach was successfully used to construct groups for the 2021 academic year. The mini-project for 2021 experienced an increase in student numbers and a higher administrative burden than the previous academic years due to the COVID-19 pandemic. The tool assisted in reducing the time and errors for group allocations.

References

1. Bäck, T., Schwefel, H.P.: An overview of evolutionary algorithms for parameter optimisation. Evol. Comput. **1**(1), 1–23 (1993)
2. Bekele, R.: Computer-assisted learner group formation based on personality traits. Ph.D. thesis, University of Hamburg, January 2006
3. Christodoulopoulos, C.E., Papanikolaou, K.A.: A group formation tool in an e-learning context. In: 19th IEEE International Conference on Tools with Artificial Intelligence (ICTAI 2007), vol. 2, pp. 117–123 (2007)
4. Graf, S., Bekele, R.: Forming heterogeneous groups for intelligent collaborative learning systems with ant colony optimization. In: Ikeda, M., Ashley, K.D., Chan, T.-W. (eds.) Intelligent Tutoring Systems, vol. 4053, pp. 217–226. Springer, Heidelberg (2006). https://doi.org/10.1007/11774303_22
5. Hou, E.S., Ansari, N., Ren, H.: A genetic algorithm for multiprocessor scheduling. IEEE Trans. Parallel Distrib. Syst. **5**(2), 113–120 (1994)
6. Hwang, G.J., Yin, P.Y., Hwang, C.W., Tsai, C.C.: An enhanced genetic approach to composing cooperative learning groups for multiple grouping criteria. Educ. Technol. Soc. **11**, 148–167 (2008)
7. Ingham, A.G., Levinger, G., Graves, J., Peckham, V.: The Ringelmann effect: studies of group size and group performance. J. Exp. Soc. Psychol. **10**(4), 371–384 (1974)
8. Koppenhaver, G.D., Shrader, C.B.: Structuring the classroom for performance: cooperative learning with instructor-assigned teams. Decis. Sci. J. Innov. Educ. **1**(1), 1–21 (2003)
9. Kulaksız, A.A., Akkaya, R.: A genetic algorithm optimised ANN-based MPPT algorithm for a stand-alone PV system with induction motor drive. Sol. Energy **86**(9), 2366–2375 (2012)
10. Lin, Y.T., Huang, Y.M., Cheng, S.C.: An automatic group composition system for composing collaborative learning groups using enhanced particle swarm optimization. Comput. Educ. **55**(4), 1483–1493 (2010)
11. Marshall, L., Pieterse, V., Thompson, L., Venter, D.M.: Exploration of partici-pation in student software engineering teams. ACM Trans. Comput. Educ. (TOCE) **16**(2), 1–38 (2016)
12. Moreno, J., Ovalle, D.A., Vicari, R.M.: A genetic algorithm approach for group formation in collaborative learning considering multiple student characteristics. Comput. Educ. **58**(1), 560–569 (2012)
13. Nand, R., Sharma, A., Reddy, K.: Skill-based group allocation of students for project-based learning courses using genetic algorithm: weighted penalty model. In: 2018 IEEE International Conference on Teaching, Assessment, and Learning for Engineering (TALE), pp. 394–400. IEEE (2018)
14. Paliwal, A., et al.: Reinforced genetic algorithm learning for optimising computation graphs. In: International Conference on Learning Representations (2020)
15. Pieterse, V., Thompson, L.: Academic alignment to reduce the presence of 'social loafers' and 'diligent isolates' in student teams. Teach. High. Educ. **15**(4), 355–367 (2010)
16. Pieterse, V., Thompson, L., Marshall, L.: Rocking the boat. In: Proceedings of the Southern African Computer Lecturers' Association (SACLA 2011) (2011)

17. Pieterse, V., Thompson, L., Marshall, L., Venter, D.M.: An intensive software engineering learning experience. In: Proceedings of Second Computer Science Education Research Conference, pp. 47–54 (2012)
18. Pieterse, V., Thompson, L., Marshall, L., Venter, D.M.: Participation patterns in student teams. In: Proceedings of the 43rd ACM Technical Symposium on Computer Science Education, pp. 265–270 (2012)
19. Shukla, A., Pandey, H.M., Mehrotra, D.: Comparative review of selection techniques in genetic algorithm. In: 2015 International Conference on Futuristic Trends on Computational Analysis and Knowledge Management (ABLAZE), pp. 515–519. IEEE (2015)
20. Sivanandam, S.N., Deepa, S.N.: Genetic algorithms. In: Sivanandam, S.N., Deepa, S.N. (eds.) Introduction to Genetic Algorithms, pp. 15–37. Springer, Heidelberg (2008). https://doi.org/10.1007/978-3-540-73190-0_2
21. Testnet. https://en.bitcoin.it/wiki/Testnet. Accessed 12 June 2021
22. Tuckman, B.W.: Developmental sequence in small groups. Psychol. Bull. **63**(6), 384 (1965)
23. Unal, M., Onat, M., Demetgul, M., Kucuk, H.: Fault diagnosis of rolling bearings using a genetic algorithm optimised neural network. Measurement **58**, 187–196 (2014)
24. Wang, D.Y., Lin, S.S., Sun, C.T.: DIANA: a computer-supported heterogeneous grouping system for teachers to conduct successful small learning groups. Comput. Hum. Behav. **23**(4), 1997–2010 (2007)
25. Yannibelli, V., Amandi, A.: A deterministic crowding evolutionary algorithm to form learning teams in a collaborative learning context. Expert Syst. Appl. **39**(10), 8584–8592 (2012)
26. Zheng, Z., Pinkwart, N.: A discrete particle swarm optimisation approach to compose heterogeneous learning groups. In: 2014 IEEE 14th International Conference on Advanced Learning Technologies, pp. 49–51. IEEE (2014)

Teaching in a Time of Uncertainty – A Practical Guide

Geoffrey Dick(✉)

St. John's University, Queens, NY 11439, USA
gfdick@aol.com

Abstract. In March of 2020, many of us were required to move our classes to an online mode – often with very little notice. For many faculty and students this was the first time we found ourselves in a learning environment that was not only unexpected but for many, not what they wanted. Special measures were called for, and unfortunately, it seems very possible we may need to continue to move between face-to-face and online classes as the pandemic ebbs and flows. Drawing on many years of diverse experience in online teaching, this paper provides some guidance for those inexperienced in online teaching in regards to the pedagogical changes that are necessary – what can work and what is likely to be problematic. The paper is written in two sections – the first is focused on using the Learning Management System to facilitate a quick move to online learning and the second part covers some of the longer-term difficulties that should be considered as we progress or move to the new environment.

Keywords: Online classes · Pedagogy · LMS · Online assessment · Pandemic · covid19

1 Introduction

Like many of my colleagues all around the world in March of 2020, I had to move my current classes in New York City online – with less than a week's notice. I consider that I know quite a bit about online classes – I have been teaching that way for 20 years on at least six different Learning Management Systems (LMS) platforms, in countries as far apart as Mexico and Malaysia, from Australia to the US and Singapore to Norway; teaching students at all levels from those starting their university studies up to those finishing graduate programs, and taught in at least 6 different major topic areas. But this was different – in the past, almost all of my online students had been there because they chose to be, this time they were there because they were given no choice, the pandemic ruled. And, to compound the difficulty, many of us were well aware that the current need for urgent action was at least possible to be short-lived as university administrations searched for ways to bring students back to the classroom. Now, as positive tests rise and subside in many countries, the future of many online classes remains in limbo. This inevitably raised the question "How much effort should I put into what may be a stop-gap solution?" In trying to relate my experiences over those twenty years to that time I learned some things. It is those "somethings" that I would like to share with you, particularly those of you who are new to online teaching or have only limited experience.

The paper looks first at a central element in online classes – the LMS and the underpinning technology. The second part of the paper considers some issues that instructors might like to consider after getting started.

2 Setting Up the LMS

The role of the LMS has been much studied [1] but its role changed significantly in March 2020. Rather than being a repository for material and supplementing the classroom, it assumed the public face of classes around the world. For most face-to-face classes we tend to use a folder structure with separate compartments for course documents such as a syllabus or outline, another for presentations, another for quizzes and exams, another for readings, etc., and we expect the students to extract the relevant or require information from these folders as they progress through the course. From a pedagogical viewpoint, this works well (in my opinion) as it gives the impression of the whole course being a composite artifact with all parts inter-relating and building on each other. For most of my online classes, I have used this structure, in the face of opposition from the instructional designers. Instructional designers [2] (among others) have argued that a better approach for facilitating learner control in the online environment is for the course to be structured into modules each containing everything that is needed for work in the course in a particular time period – usually a week. The point about user control is a good one in the current environment where we are encouraging students to engage.

For my classes this Spring, I changed my mind, not for pedagogical reasons, but for enabling the students to feel more in control and cope better with the exigencies of online classes, into which they had been drafted. It is important, I think, to remember that most of them didn't sign up for this. I once wrote a paper claiming that online classes were not for all of the people, all of the time [3]. In these last few months, most of us have been thrust into a situation where online classes are for all of the people (professors and students alike) and for all of the time – or at least for most of last semester and probably, for some at least, well into the future. Many students were not well equipped to adapt to an environment that required them to be independent learners with the underlying necessary characteristics of being motivated, disciplined, and good at time management.

A survey I conducted at the end of the last term included an open-ended question along the lines of "what did you like about this course?" Many students commented on the module arrangement – these two quotes reflect the general themes expressed:

> "I like how once we were fully online, that you created modules for completion which made it a lot easier to see the work we had to do", and

> "The "modules" that you uploaded that clearly told us which assignments where due on what date exactly, AND that we can see weeks ahead so if something were to come up we have time to do the assignments early and not miss out."

Figure 1 below provides a screenshot showing the module approach from a recent course of mine.

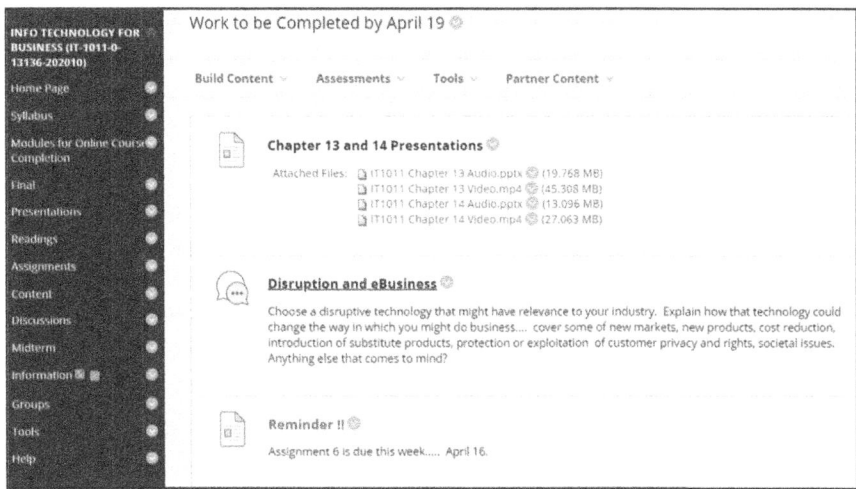

Fig. 1. A learning module

We need to bear in mind that students usually take several classes and each professor does things differently (a version of "academic freedom" perhaps?) – putting everything to be done in one place does make it easier for students to complete the work due.

One step that could be taken then, is to set up the LMS in modules – they will still work satisfactorily if the students come back to the classroom, and are probably better for online, should it continue. However, whether you go with the more traditional approach or the module approach, here are some guidelines to aid you in setting up your class, assuming you are largely left to your own devices:

a) Don't get hung up on the technology. In other words, it has to be there, but it won't make your course successful by itself. Remember that many students have limited resources and that most technology has a learning curve. Also, at least for now, students are likely to be taking several classes online and it is generally seen as beneficial if they don't have to deal with several different ways in which to participate. Try and keep it simple for them.
b) Remember that many of your students will not have access to the software and fast, reliable, internet connection that you do – you do not want them to become disillusioned with your course because they are unable to download a file or stream a video. As students vie with other members of the family for computer resources they will try to cope by using their 'phones. In most cases, small file sizes for formatting video or audio are quite acceptable, particularly when viewed on a small screen. Test it out – there is a big difference in download time, often with no discernable difference in quality. It is important that you try and discern what communications difficulties and costs your students may be experiencing in the operational domain and adjust your expectations and demands accordingly. This is particularly so in countries where there may be widespread access issues but almost all of my classes have had some students who have similar problems at the individual level.

c) Always make sure you can see the LMS the way the students see it. Most allow a "student view", but these do not always cover everything; some allow dummy students (in my view the best possible approach but increasingly being phased out, apparently due to copyright and fraudulent use issues ... I am skeptical). I ask a student to log on and share screens with me, at least for the initial set-up. Getting students to help like this is beneficial in another way too – it promotes student engagement and lets them see that you care about doing things right.
d) Keep the pages uncluttered and reserve the opening page for announcements and a short menu.
e) Remember that at this time many students will be under severe stress – and dealing with several different classes, perhaps offered in different ways. Find a way to ensure that students can easily determine if there is anything to which they need to be paying attention – "Announcements" that appear on the home page for the course (in the student view) are good but students need to be able to readily discern if there is anything new – I use a colour-coding system, announcements with a title in red appeared in the last few days. Some LMS allow you to show the most recent announcements at the top of a home page.
f) One document at the top of the menu should be something like "Read this first" and it should provide a brief and succinct overview of the steps to be taken to get started and how best to contact you. You might even consider a short video of a screen capture with you moving through the LMS in the student view.
g) Decide if you would like to run "office hours". If so make it clear in the "Read this first" document how this will work. You have many options here from a WebEx meeting room to a chat session or even a discussion board. Typically, I do not run office hours as such, (apart from one or two sessions early in the course, when many students are unsure about what is expected of them) but I do mostly respond to messages or emails very quickly, even if just to say "I will get back to you". If you don't want to respond promptly to messages, "office hours" are essential.
h) Decide on the traditional or module approach and stick to it – it will cause great confusion if you change horses midstream.
i) Decide on how you would like to make material available. I am an advocate of the principle that apart from the textbook, everything needed for the course (apart from outside research) should be available on or through the LMS. Pdfs, video links, quizzes, assignment or project submissions, external readings, library links etc., all fall into the category of "everything". Some like to use a shared drive – but I believe the one-stop-shop makes it simpler and easier for them to cope.
j) Many LMS provide extensive statistics for analysis related to your course – who has watched your presentations and for how long, how many times students have accessed certain sections, grading statistics etc. They can be a useful tool in identifying underperforming or "at-risk" students – an essential task in online classes with involuntary enrolment. Make sure you know what is possible (some of these have to be set up in advance) and decide if you would like to use them. A simpler way could be to set a short quiz each week and monitor completion – a great "early warning" system for identifying students likely to fall into trouble.
k) Try and stay at least a couple of weeks ahead, ideally set up all of the main items (presentations, assignments, readings) at the beginning of term. It does not create a

good impression to be loading material a day or so before you expect the students to be working on it. On the other hand, there is really no need to put all the presentations up at the beginning – of it can be seen that a place for them is available, that will be sufficient. This may save some effort should we return to the classroom mid-semester.
l) If you make an amendment to the syllabus, course outline or schedule, draw attention to it as an Announcement. If it is important, send an email as well.

3 Getting the Class Moving

Once we have the LMS and technology in place there are a few more central issues that we can address relatively quickly and take action to promote student engagement and learning.

3.1 Discussions

Discussions on an LMS platform are intended to provide a forum to replicate in-class discussion. But do not look to the discussion board to provide a substitute for the sort of question you might ask during a presentation to ensure that the students are following your drift. The synchronicity doesn't work. On the other hand, in my experience they can provide a useful base for developing ideas further and having the students build on each other in the development of ideas or comprehension – this is probably the best part of the course for the students to learn from each other. However, there is at least some evidence that discussions do not contribute much to course satisfaction and student achievement [4].

As an aside the discussion board can provide a useful resting place for Frequently Asked Questions – have the students post their questions there along with your answers, although in my experience this works well only for the first couple of weeks. After that, they revert to asking directly, rather than bothering to look at the FAQs.

To use the discussion board effectively:

a) You must be prepared to comment frequently on the student postings. If you do not do so, it will not take them long to assume you are not interested in what they have to say. For my small class last Spring I tried to make some short comments on each student's post, for my larger one I put together a summary at the end of the discussion period, where I mentioned several contributors by name and chose the odd quote made by one or more to illustrate a point. Feedback to the students, making them feel they are appreciated is very important in times of stress.
b) For large classes (say more than 20) it is probably impractical to expect every student to read every post. And dozens of posts and comments on posts are difficult for you to stay on top of too. I like to break the class into groups of about 8 and appoint one person for each topic to write a summary of the postings on that topic at the end of the discussion period (for grade points). It can be a nice review exercise to have the students compare the summaries across groups.
c) In order to ensure postings keep coming, you will almost certainly have to use the heavy hand of grading – (and here is the difficult bit) in a way that reflects student

effort but does not necessarily provide them with a "checklist" of things they have to do to get full points – post by a particular date, comment on the posts of two others, write 100 words etc. I usually try and provide individual feedback about halfway through the term as to the value I have seen in their postings to date – the key point to get across is that you want to see they have been thinking about the issues, rather than checking boxes.

d) If your LMS allows it, think about the desirability of not allowing students to see other posts on a topic before they make their initial one.

e) Shut off the discussions at the end of the discussion period and do not give credit for late submissions – no one is going to go back and reread old discussions. Point out that these are not assignments or projects, they are in fact "discussions".

3.2 Group Work

Group work is particularly problematic in online classes, although in the Spring of 2020 at least all of my students had had some contact with each other. Many students detest it as they feel it only adds an extra degree of difficulty to a task. Communicating with people they don't know, organizing work allocation, subsequent workflows and task completion, and carrying non-performing team members are all potential frustrations, particularly in times of external stress. On the other hand, we want the students to experience group work as it enables more meaningful tasks to be set and assessed, we need to demonstrate to the accrediting agencies and potential employers of our students that they are equipped to work in teams and it gives them experience in working in spatially- and time-distant environments – surely what they will experience in the workplace. And it can reduce our grading load.

My suggestions for group work include:

a) Keep it to a minimum. This clearly depends on the learning objectives for the course and what you are trying to achieve, but I usually don't set more than one significant group assignment as part of the course assessment. Group work is good to share – a team that has not performed well is seen as a team, not individuals, and this, in turn, can raise group performance, rather than disillusioning one student.

b) In allocating students to teams in online classes we usually have little choice but to do it randomly. For some projects it makes sense for the students to identify their strengths and skills and be allocated so that these are spread out among teams – I would encourage that if you can. Another possible allocation route is to ask all the students to introduce themselves and have them use this to pick their teammates. Still, some are likely to get left behind.

c) I usually form groups at the beginning of the semester and set tasks, discussion board contributions etc., within those groups. Such tasks can be quite small (a short paper or video presentation) but provide a chance to "get to know" each other and see each other's work.

d) Most LMS allow for multiple submissions of student work – you will probably save yourself a lot of hassles if you allow groups to make several submissions. Most students leave work (or least some of it) to the deadline and confusion often arises

over last-minute edits and changes. I allow 2 or 3 submissions, and tell them this, but also tell them I only consider the last submitted.
e) For the assessment of group work I include in my syllabus, and draw attention to it in the wording of the assessment task, that I retain the right to conduct a peer review where the group members assess each other's contribution and amend the grades accordingly. Over the years, I have not had to carry out this threat very often.
f) Try and set a date, several weeks before the due date, by which at least one member of each group has to report on progress and work allocation – hopefully, this will provide an early warning of problems that might be encountered.

3.3 Ethics

Somehow, I am more likely to see students as individuals in an online class as opposed to a "student body" in the classroom. In addition, while the resources available in the classroom and on campus are there for all to use, that is clearly not the same for those taking a class off-campus. Some students will have a quiet, private place to work at home, with a high-speed, reliable, internet connection, a modern laptop and printer and all the software that might be useful. Others will have none of these things and try and do their work on outdated equipment, or in the local library, or on their'phone. Again these issues take on particular importance in circumstances such as those we are facing now.

Accordingly, I worry constantly about the ethics that surround my online teaching. Because I am focused on establishing a rapport with the students and treating them as individuals who matter, I find it very easy to commit ethical transgressions such as affording special consideration (such as late acceptance of work) to one and not others, taking circumstances (inability to watch a video for example) into account for one but not know if similar circumstances apply to others, developing an expectation that all students should have access to appropriate technology because I do, or even taking personal requests into account more than I would normally.

My suggestions here are first, be aware of what you are doing and all the decisions you take; secondly, document what you are doing and why you took the decision you did – the very need to document an action makes us think about its acceptability; and thirdly, discuss how you run your class and these kinds of issues with your colleagues and assess whether what you are doing is in keeping with their practice.

3.4 Testing and Assessment

This is too complex an area to be covered in detail here, but a few lessons I have learned over the years might be helpful to some. If you are fortunate enough to have a small class, your knowledge of the students' work will provide you with good guidance as to the grade well ahead of any final. Should this be out of line with a test performance that may ring some alarm bells? The second part of the solution is to set exams that are similar to what we know as "take-home exams" where we ask the students to apply course concepts to given scenarios or cases. The downside of course is that these are more difficult to set and time-consuming to grade. A colleague of mine uses case presentations produced as a major assignment during the term, makes them available to the students on the

LMS for the exam and asks them to choose one and evaluate it in accordance with the course material – I think I will try this. Other ways to conduct assessment include, at the beginning of a test, drawing students' attention to the penalties for cheating, looking for wording or phrases that don't sound like the student's expression and checking for them online, using proctored exams where possible, setting short time limits for multiple-choice questions, random delivery and questions sets. None of these is foolproof and I believe that this area remains a challenge for the acceptance of online classes.

4 In Conclusion

The way you run your online classes will probably largely be up to you. Remember that most students at the undergraduate level did not anticipate taking online classes when they enrolled – at least not as many as they are facing now. In the end-of-term survey, I ran at the end of Spring, 2020, less than 10% of the students did not "agree" or "strongly agree" with the statement "I really hope we are back on campus next term". Although the analysis of this survey data is ongoing as part of a larger study, it seems that this desire was largely driven by satisfaction with the courses, which in turn was based on a preferred way of learning and the self-efficacy necessary to do so. Specific problems were reported in the areas of a room to study in, broadband, time management, and access to hardware and software. Not unexpected, perhaps.

I am hopeful that the issues raised above have gone some way to mitigate some of the concerns reported by the students in that survey – things like a standard platform and non-complex technology, regular feedback and being available when they have a problem, organized workflow and task completion, easily accessible course materials, and attempts to minimize stress. It is important to remember that the face-to-face environment provides some level of inherent discipline which is largely absent online. Somehow, if we want our students to succeed, we need to find ways to keep those dedicated and inquiring minds on board. What I have outlined above are modifications to procedures that we might otherwise introduce to make the task a little more manageable for them, and indeed, manageable for us. None of the suggestions I believe, are onerous and can easily be transferred back into the face-to-face teaching environment. I hope that you will find some of it useful.

References

1. Eom, S.B.: Effects of LMS, self-efficacy, and self-regulated learning on LMS effectiveness in business education. J. Int. Educ. Bus. **5**(2), 129–144 (2012)
2. Stansfield, M., Mclellan, E., Connolly, T.: Enhancing student performance in online learning and traditional face-to-face class delivery. J. Inf. Technol. Educ. Res. **3,** 173–188 (2004)
3. Dick, G.N.: Teaching online: what price student satisfaction? In: AMCIS 2003 Proceedings. p.399 (2003). http://aisel.aisnet.org/amcis2003/399
4. Cho, M.H., Tobias, S.: Should instructors require discussion in online courses? Effects of online discussion on community of inquiry, learner time, satisfaction, and achievement. Int. Rev. Res. Open Distrib. Learn. **17**(2) (2016)

Utilizing Computational Thinking in Programming to Reduce Academic Dishonesty and Promote Decolonisation

Suné van der Linde(✉) and Janet Liebenberg

School of Computer Science and Information Systems, North-West University, Potchefstroom, South Africa
{sune.vanderlinde,janet.liebenberg}@nwu.ac.za

Abstract. Higher education in South Africa has been in the spotlight in the past few years with calls for decolonisation of the curriculum and other matters. We teach a first-year programming module that is challenging to decolonise since the origin of programming languages is inherently Western. Students often do not resonate with some examples used, let alone abstract concepts of programming in general. During COVID-19, emergency remote teaching and learning were adopted and we had to be mindful of various limitations, such as data usage and bandwidth. We experienced difficulty expanding each student's frame of reference. Furthermore, increased academic dishonesty occurrences were encountered. This paper focuses on contextualising the module content, promoting computational thinking, and reducing academic dishonesty. This was achieved in an action research cycle through enriching our assessment practices by creating a weekly assignment where the principles of computational thinking were applied within a problem-solving learning environment. It was found that most students had positive perceptions about the intervention and their views and experiences are reported.

Keywords: Assessment in programming · Academic misconduct · Decolonisation · Introductory programming · Computational thinking · Assessment · Action research

1 Introduction

Calls for decolonisation in education in South Africa have become prominent [9, 16, 20, 28]. Programming languages and textbooks are predominantly written in English with Western examples depicting measurement systems and currencies unfamiliar in the South African context. Decolonising the programming curriculum is already a challenge in classrooms; therefore even more so during remote emergency teaching and learning.

During remote emergency teaching and learning, we faced a new world of teaching programming to students who had enrolled for face-to-face teaching and learning. We found ourselves in a situation where videos only covered concepts because of limitations, such as data and bandwidth. We had less opportunity to broaden the frame of reference for our students as we normally would in class. An alarming rate of academic dishonesty was also recorded which endangers the integrity of our assessments and the quality of our degree.

In this paper, we explore and report on students' perceptions about an intervention that we used to allow students to bring their context to the learning environment, to promote computational thinking and to reduce academic dishonesty. We use computational thinking as a foundation for our instructional design and interventions planned within a programming module.

2 Conceptual Framework

2.1 Computational Thinking and Programming

Computational thinking is an approach used to solve problems using specific concepts, such as abstraction, recursion and iteration. This approach is used to create artefacts across disciplines through the processing and analysis of data [2, 14]. Computational thinking is defined as a conceptual foundation necessary to solve problems effectively and reusable in different contexts [31]. Berland and Wilensky [5] posit that computational thinking embodies the idea of solving problems using a computer. In this study, we resonate most with the following definition of González [11:2438], since the link with computer science concepts is evident here: *"CT involves the ability to formulate and solve problems by relying on the fundamental concepts of computing, and using logic-syntax of programming languages: basic sequences, loops, iteration, conditionals, functions and variables."*.

Five core skills of computational thinking are applicable within computer science, namely problem solving, building algorithms, debugging, simulation and socialising [15]. In a review of 45 computational thinking research papers, the most common components or skills of computational thinking are decomposition, abstraction, algorithms and debugging, with the addition of iteration and generalisation [30].

2.2 A Problem-Solving Learning Environment (PSLE)

Lye and Koh [19] developed a computational thinking instructional design based on the review of 27 computational thinking papers. In their review of papers, 20 out of the 27 papers were within the context of higher education. The instructional design essentially promotes a PSLE in order to foster computational thinking skills. The envisioned environment (PSLE) should expose students to authentic problems, activities based on cognitive constructivism, scaffolding and reflection. A PSLE is a comprehensive strategy based on the views of many leading authors in the field (Table 1).

2.3 Decolonisation of the Programming Curriculum

Calls for the decolonisation of education in South Africa are prominent [9, 16, 20, 28]. The institution where this study was conducted also made explicit statements regarding the consideration of the transformation agenda in teaching, learning and assessment matters [25]. Furthermore, the institution's stance in terms of teaching and learning is that

Table 1. PSLE components for instructional design, adapted from Lye and Koh [19]

PSLE component	Description
Authentic problems	Problems should be set within a context with regard to the students. Students tend to be more engaged on an intellectual level in the learning process when the problem is relevant to them
Information Processing	Computational concepts are acquired through information-processing techniques that focus on mental model constructions (constructivism)
Scaffolding	Scaffolding the program construction into smaller more manageable tasks. This step is based on the constructionism process where meaningful products are built for themselves or others as a result of continuous knowledge construction
Reflection	Students reflect on computational processes and their programming process, either by self-reflection or peer-reflection

decolonised teaching and learning practices should *"inspire students to think critically on and engage with issues, such as discrimination, racism, inequality, poverty, colonialism, alienation, inclusion and ethical conduct. Furthermore, decolonised teaching and learning allows students to interpret curriculum content based on their own experiences, according to their cultural norms, personal belief systems, preferences and backgrounds, and to share their interpretations with fellow students as valid and valued real-life experiences."* [24:4].

Gumbo [12] posits that many educators ignore pedagogical perspectives that would allow recognition of indigenous learners in their teaching. This might lead to passive learning when learners cannot relate to the content and teaching strategies [17]. The problem worsens when curricula and learning material does not contain indigenous knowledge, or simply contains content and examples that are not always relevant to the particular students [12, 22]. In the programming classroom, textbooks are mostly from a western origin, with western examples to which not all students can relate [12, 22]. Integration of students' background knowledge and experiences is necessary to follow a culturally responsive setting in the classroom [12].

2.4 Academic Dishonesty in Programming

Academic integrity is vital for the reputation and credibility of educational institutions, regardless of their teaching modality [27]. Cheating or academic dishonesty occurs when a student engages in dishonest behaviour and gains an unfair advantage, misrepresenting his/her ability and knowledge within the course [1]. Liebenberg and Botes [18] found in their study at the same university where this study was conducted that 44.4% of the students admitted to plagiarism and 59.4% admitted to sharing their code. Academic dishonesty in programming courses usually entails students copying solutions from one another, and sometimes slightly changing names of variables in the source code, and other times not even changing anything. Students could also copy the source code of the internet and sometimes they find the repository of solutions online for the textbook in use [18].

Higher education's necessitous switch to remote teaching and learning brought about by Covid-19 precipitated many anxieties about education in a fully remote environment. In his reflection on academic integrity during Covid-19, Scurr [29] highlighted four features: 1) a belief that cheating would automatically increase in an entirely remote environment; 2) a growth in the dividing perspectives on how to deal with or penalise academic misconduct; 3) an ultra-consciousness of contract cheating and file-sharing sites; and 4) a shift in how to best validate student work.

According to Adzima [1], academic dishonesty can be caused by the lack of face-to-face sessions and the challenges that come with learning in an online environment. Some of these motivations include the lack of personal connection between student and lecturer [21].

3 Research Design

In this study, action research (AR) is used as a tool to enquire and bring about social change through the improved understanding of social practice (teaching and learning to program), as well as to improve the practice (teaching and learning to program) itself [3, 4, 13]. AR is an iterative process that improves understanding of the problem with each iteration and includes the phases illustrated in Fig. 1:

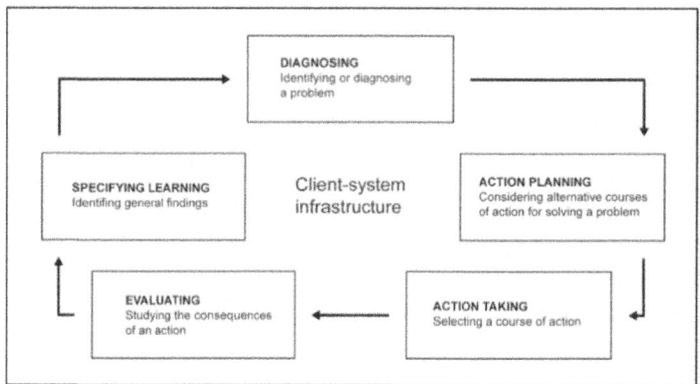

Fig. 1. The action research cycle [8]

In this paper, a specific intervention used in one AR cycle (cycle 8) of a longitudinal study focussing on using critical systems to improve the programming skills of students through the explicit use of computational thinking is reported on. A PSLE grounded in computational thinking is used, but the scope of this paper is focused on a specific element of the PSLE, namely authentic problems (Table 1). In terms of the evaluation phase of AR, students' feedback regarding action taking were gathered, using a mixed methods approach. Tashakkori and Creswell [32:4] describe mixed methods as: *"Research in which the investigator collects and analyses data, integrates the findings, and draws inferences using both qualitative and quantitative approaches or methods in a single study or programme of inquiry*". Creswell and Clark [6] suggested four major types of mixed methods design: (1) triangulation; (2) embedded; (3) explanatory; and (4) exploratory. In this study, the type of mixed methods research was explanatory, as the objective of the qualitative investigation was to supplement the quantitative investigation and to better understand and explain the observations of the quantitative investigation. For the evaluation phase of AR, the quantitative data was collected through the use of a survey and the qualitative data was acquired through the comments of survey respondents.

In the next section the steps followed during action research are covered.

4 Action Research

This section is discussed in terms of the AR phases for AR cycle eight.

4.1 Diagnosis

The evaluation and specify learning phases of AR cycle seven is important to inform the diagnosis of AR cycle eight. In AR cycle seven, an emergency remote teaching and learning approach was followed in response to national lockdowns in South Africa due to the COVID-19 pandemic. The basic instructional design followed in AR cycle seven included completing the required reading, watching a tutorial video, completing a concept quiz for learning, completing two to three textbook programming exercises (Fig. 2), and reflection upon reaching the outcomes.

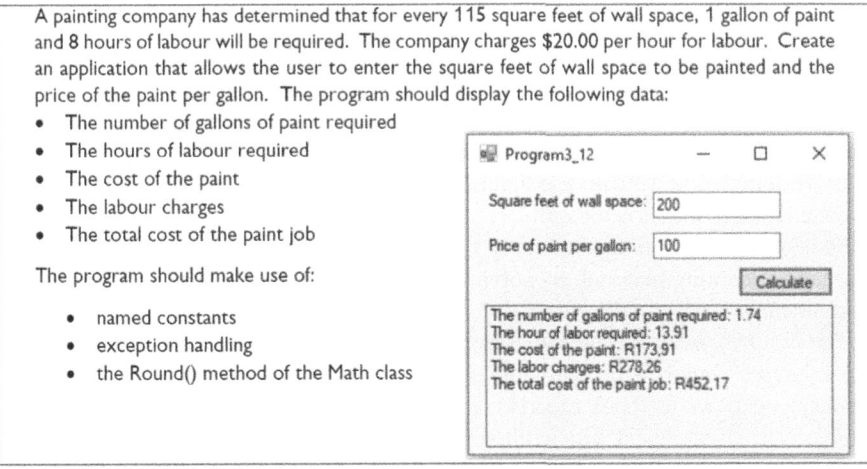

Fig. 2. Textbook exercise example

The programming exercises used are out of the prescribed textbook which can be viewed as a Western textbook with examples that assume a certain background. The reality is that South African students studying at the North-West University have very diverse backgrounds. As an example, the original text in Fig. 2 uses terms, such as feet, gallons, and dollars – unfamiliar and outlandish terms for our students.

After reflecting on the findings of AR cycle seven, two main problems were identified with the current approach:

An Alarming Number of Academic Misconduct and Dishonesty Occurrences. Moving to remote online teaching and learning during COVID-19 brought challenges for all involved. Students were overwhelmed, and it seemed as if they were completing assignments in a panic mode, working from one deadline to the next, which might have hindered deep learning. We found a major increase in academic misconduct and dishonesty.

Students Did Not Have Enough Opportunities to Bring Their Context Into the Learning Environment. Incorporating authentic and relatable problems has been a challenge. In previous AR cycles, we incorporated one group project that allowed for social interaction at the end of the semester with a variety of possible scenarios. During the national lockdown in South Africa, we found it too problematic to attempt giving a group project for this module, because of connectivity issues, issues with demonstrations of the projects, and the trustworthiness of the projects.

4.2 Action Planning and Action Taking

The two lecturers realised that they would have to rethink assignments and assessment. The planning process was in the form of telephone conversations where possibilities of how to intervene in order to lower the number of academic dishonesty occurrences within the framework of computational thinking were discussed.

One lecturer mentioned that the only way to address academic dishonesty was to have all students work on different assignments weekly and it was decided that a weekly project (referred to as a creative assignment) where students are provided with a rubric, but not a scenario might work well.

Students were encouraged to apply knowledge learned through concept videos and other online learning material, by solving problems within their local community and environment.

The first two weeks of the semester students received the traditional "guided textbook exercises", but then the instructional design was amended to where students would complete one to two textbook exercises with specific instructions, and in addition, complete a creative assignment with a rubric (Fig. 3) over a period of nine more weeks until the end of the semester.

The aim of this assignment is for you to apply higher order thinking skills to showcase your ability to reach the outcomes of this study unit. This assignment is project based. This means that you'll receive some minimum requirements for developing your own application. You'll need to come up with a concept, design the graphical user interface, and write the code.

1. Project_SU4_1

	MINIMUM REQUIREMENTS	MARK
INPUT	Three (3) variables in total	1
	One (1) constant variable	
PROCESSING	One (1) while-loop	2
	Validate at least one (1) input using TryParse (including else statements with informative dialogues regarding invalid input)	1
OUTPUT	Format output accordingly, displayed neatly	1
CONTROLS	Appropriate controls used for input	1
	One (1) CheckBox control	1
OTHER	Apply simple exception handling	1
	Neat and organised appearance	1
	Compile and Run	1
	TOTAL	10

Fig. 3. Creative assignment example

4.3 Evaluating and Specifying Learning

4.3.1 Participants

The programming course that forms part of this study is User Interface Programming 1 within the BSc in Information Technology degree. This course utilizes the C# programming language to achieve the outcomes. It is usually presented by two lecturers at two different campuses of the North-West University (NWU), situated in Potchefstroom and Vanderbijlpark which are about 120 kms apart in South Africa. In addition, the course is offered as a distance-learning course. The module content, assessments and contact time are aligned.

In the second semester of 2020, the course was offered as an emergency remote online course because of Covid-19 and the 632 students were taught by two lecturers. The students of the Potchefstroom campus and the distance students were taught by the one lecturer and the remaining students were taught by the lecturer on the Vanderbijlpark campus.

4.3.2 Instrument, Data Collection and Analysis

In this study, a questionnaire regarding the creative assignments of the course was developed. Students entered their biographic data in the first section. Six questions (Table 2) were developed for section two, of which four focussed on the students' perceptions on whether the completion of creative activities reduced plagiarism, empowered students to solve problems within a community, promoted higher-order thinking skills, and improved performance. Two questions focussed on whether students relate with, and comprehend textbook scenarios. The questions in the second section were accompanied by a five-point Likert response scale from 1 (Strongly disagree) to 5 (Strongly agree). Lastly, two open-ended questions asked: *"What did you like about completing creative projects? (where you choose your own scenario)"* and *"What didn't you like about completing creative projects?"*.

After the final assessment, but before the final marks were released, a link to the anonymous online questionnaire was sent via the e-learning system to the students taking the course. 509 usable responses to the online questionnaire were received, indicating an overall response rate of 80.5%.

Table 2. Profile of respondents (n = 509)

Criteria	Categories	Number (%) of students
Group	Potchefstroom	283 (55.6%)
	Vanderbijlpark	160 (31.4%)
	Distance	66 (13%)
Gender	Male	352 (69.2%)
	Female	157 (30.8%)
Ethnic background	African/Black	251 (49.3%)
	White	215 (42.2%)
	Coloured	14 (2.8%)
	Indian/Asian	13 (2.6%)
	I don't wish to say	16 (3.1%)
Academic performance	75%–100%	213 (41.8%)
	60%–74%	160 (31.5%)
	50%–59%	79 (15.5%)
	Below 50%	57 (11.2%)
I mainly used the following device to write my C# programs	Desktop computer	74 (14.5%)
	Laptop computer	411 (80.7%)
	Minicomputer (Netbook)	8 (1.6%)
	Mobile phone with Internet access	16 (3.1%)
Wi-Fi access at home	Yes	286 (56.2%)
	No	223 (43.8%)

For this course, the gender profile is typical of most programming classes with only 30.8% of the respondents being women. Based on the ethnic background, it is clear that it is a diverse group of students with distinct cultural norms, personal belief systems, preferences and backgrounds. Within the African/Black cohort, the students on the Vanderbijlpark Campus are mainly Sesotho-speaking, whereas the Potchefstroom students are Setswana-speaking [23].

It was reported by 41.8% of these students that they expect to obtain a distinction for this course. Although students had to endure emergency remote teaching and learning, the self-reported academic performance of the students had increased in comparison with previous years – this phenomenon can probably be attributed to academic dishonesty and not necessarily improved learning [7]. More than 80% of the students used a laptop to write their programs and 43.8% reported that they do not have wireless Internet access at home – by June 2020 the NWU had spent approximately R22 million on devices and data for its students during lockdown [26].

Analysis of the quantitative data (Biographical info and six Likert-scale questions) was done in SPSS Statistics 27. Frequencies were calculated and basic analysis was done by calculating the mean values and standard deviation of each of the items. For the analysis of the qualitative data (Two open-ended questions), ATLAS.ti 9.0 was used. Content analysis was conducted on all responses open-ended questions and thematic analysis was utilized to categorise responses. The data was coded according to six themes that were derived from the six questionnaire questions (Table 3) as well as the themes that emerged during the content analysis. Since the product of qualitative research is richly descriptive, some results are presented in the form of quotes from the participant comments.

5 Results and Discussion

5.1 Quantitative Results

The statistical results for section one (Biographical info) and section two (six questions) of the questionnaire are summarised in Table 3 and Fig. 4. Table 3 shows that the mean values of all six items in the questionnaire are relatively high (with 3.0 being neutral).

Table 3. Items and descriptive statistics (n = 509)

Criteria	Mean*	St. dev.
Plagiarising during the completion of creative activities is harder than completing a textbook activity where the scenario is provided. (Plagiarism)	3.78	0.950
I always relate to the scenarios of the programming problems in the textbook. (Relate)	3.55	0.994
I always understand the scenarios of the programming problems in the textbook. (Comprehension)	3.73	0.925
Since I could choose my own scenario for my creative project, I feel empowered to solve problems within my local community. (Empowered)	3.69	1.078
My higher-order thinking skills were developed by completing creative projects. (Thinking)	4.04	0.959
The completion of creative projects helped me to improve my performance in assessments (tests). (Achievement)	3.83	1.078

*Likert-style responses were ranked from 1 to 5 respectively

It is clear from Fig. 4 that the students in this study had an overall positive experience with the creative assignments since the majority of participants agreed and strongly agreed with the six questions in Sect. 2.

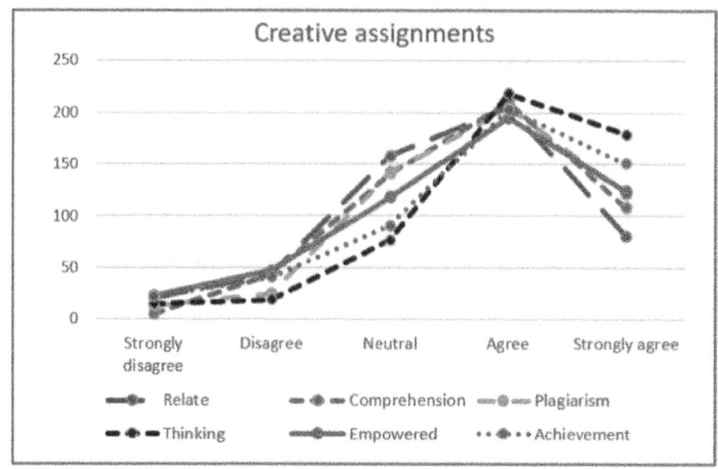

Fig. 4. Frequency of selections

In this following section, the quantitative data presented above, in conjunction with the qualitative data relating to the question "*What did you like about completing creative projects? (where you choose your own scenario)*" will be discussed.

5.2 Qualitative Results – "Likes"

Plagiarism and academic misconduct
The students reported in the questionnaire that the completion of creative activities reduced the potential for plagiarism and one of the students commented:

> It's brilliant. I had my initial concerns with plagiarism after last semester's CMPG111. I hypothesised that there would be many false positives since we all used the same logic provided in the textbooks to solve the same problems that we were presented with. With the "creative" approach it cleverly mitigates this problem. The probability of students submitting the exact same project is significantly lower. I consider it a win for both the student and the lecturer. Thank you.

Interestingly enough, one student commented in the dislikes part of the questionnaire:

> *The fact that it was hard to copy anyone else's work.*

Relate
The question with the lowest mean was if the students always relate to the scenarios of the programming problems in the textbook. However, the students were extremely

positive about how the creative assignments allowed them to relate to their real-world experiences.

> *I could integrate my hobbies for example Formula 1 into programming that was very fun. The places where my mind went with programming is mind-boggling.*
>
> *They stimulated a lot of creation from my own mind and hands. Which opened me up to a lot of possibilities for problem solving larger issues in the world.*
>
> *That we could imagine problems we would love to solve around us.*

Comprehension

The group of students did not feel excessively optimistic about their comprehension of the scenarios of the programming problems in the textbook, but some commented on how the creative assignments aided in the comprehension of their work.

> *It forced me to make sure that I fully understand the concept. And required a lot of thinking, which is useful in the learning process.*
>
> *Solving my own complex problems helped me enhance my skills and helped with the understanding of the work.*

Empowered

The students reported that they felt quite empowered and they are able to solve problems within their local community.

> *I would feel like I am making programs that could actually be beneficial to myself and other people around me.*
>
> *I sometimes felt like I was part of a company where we needed to develop apps. Made me feel like a programmer even though I'm just a first-year student.*
>
> *I was able to self-identify where my strengths and weaknesses lie.*

Thinking and learning

The question with the highest mean in the questionnaire was if students think that their higher-order thinking skills were developed by completing creative projects. In addition, students commented on their thinking and learning:

> *My high order thinking were increased, I had to think to the core and I really enjoy tapping into my inner strength and being able to solve the projects I did not think I could solve.*

Achievement

The students reported positively through the questionnaire and open-ended question that the completion of creative projects helped them to improve their performance in assessments.

> *I got an opportunity to apply my learned coding skill to a level beyond just means to pass an exam.*

I liked that I could choose my own solution to a problem. I didn't lose marks because my coding wasn't exactly like the memo, unlike with the programs where scenarios were provided.

Creativity, freedom and thinking out of the box
For the majority of the students, the fact they liked the most was that they could be creative, think out of the box and had the freedom to express themselves.

It helped me to be a bit creative. STEM doesn't always leave room for creativity, and by creating my own projects I felt more familiar with the work and it felt like I was doing something that peeked my interest. It was never boring and made a subject that I was unfamiliar with really fun.

It feels more enjoyable to create instead of "answering". Promotes creative thinking.

I could put in my own jenesiqua to my programs and format them as far as humanly possible for me. Therefore I could express myself and my way of thinking and programming in multiple everyday scenarios.

Challenge and accomplishment
Students felt challenged and they liked the sense of accomplishment when completing the creative projects.

I liked the challenge they gave me and how they kept me on my toes. I enjoyed every project I successfully completed.

It made me feel really smart being able to program my own scenario and it working.

Nothing
About 40 of the 509 students indicated that there was nothing they liked about the creative assignments and they found them repetitive and stressful.

I did not enjoy the creative projects and would have preferred more complex problems that was given to us. In a working environment you will almost always be given a spec of something to complete.

I didn't like them at all. They felt repetitive.

Nothing, they were stressful.

In the following section, the qualitative data relating to the question "*What didn't you like about completing creative projects?*" will be discussed.

5.3 Qualitative Results – "Dislikes"

Nothing
In contrast with the 40 odd students indicating that there was nothing they liked, more than 60 students declared that there was nothing they disliked about the creative assignments.

Nothing: They were GREAT!

Nothing, i think it's a good thing that you are given the choice on how to write your assignments.

There is absolutely nothing that I did not like about creative projects.

Time

The most common complaint was that the students found the creative assignments to be time-consuming.

They are time consuming because you spend a lot of time trying to think of what to code. It would be easier if we were given projects to code.

It takes time to think of a problem and i do not see why i should be able to create a problem and not just solve one.

It always took me a long time to come up with a problem to solve with my program every week.

Difficulties

Students experienced difficulties with a number of issues.

It was difficult to come up with an idea, to create codes on your own, thinking about an idea that was both practical and which could fulfil the criteria of the project, to understand and complete some of the projects, and to think creatively under pressure.

It's when you don't know something to code and when you search it on other sources it becomes difficult because every source is not reliable.

Requirements

Students were very vocal regarding the requirements set out in the rubric (see Fig. 3) of the creative assignment.

I felt very restricted in the codes that I wrote. I'd sometimes add functions that weren't necessary but just to meet the criteria that was given, and that I'd write bad code just to appeal to the criteria.

...discouraging when you know you can, for example, generate a random number, but have no idea how to make that useful.

Achievement/Marking

It is common knowledge that students are always concerned about their marks and how their assignments are marked. The student assistants who marked the weekly assignments might have missed a few things, since all the programs they are marking look different.

I think the marking criteria has to be clarified a little more. I often found myself having to bother people that were clearly already very busy to query marks and in most cases it turned out that the marker had interpreted the requirements differently and erroneously marked me down.

> *If you implemented a concept not known to the "rubric" that does the same work as the rubric requirement you would still lose marks because it is not the way the rubric is looking for. For me it feels that it takes away creative problem solving a certain problem with "unorthodox solutions"*

No problem solving
Some students were of the view that they do not develop their problem-solving skills and that the creative assignments were superficial exercises.

> *I like solving problems, not finding the problem to what the solution is supposed to resemble.*
>
> *I did not like those projects at all, because one will never have to write such a program for a client. The client usually will come to the developer with a problem, and a desired destination/outcome.*

Feelings
Some students expressed feelings of isolation and disappointment:

> *You are on your own with no little help besides you, it is hard to create a structure specially drafted by you.*
>
> *They made me feel stupid because I'm not a creative person.*

5.4 Reflection

From the perspectives of the lecturers (also the researchers) that presented the course, the use of creative assessments was beneficial in terms of encouraging students to utilize higher-order thinking skills to solve problems according to the rubric criteria. The creativity shown in projects were encouraging, but not all students enjoyed the challenge of creating a problem and solving it. The creative assessments promoted decolonisation by encouraging students to bring their context into the classroom and it seems that they enjoyed this aspect as one participant mentions "*I would feel like I am making programs that could actually be beneficial to myself and other people around me.*". The lecturers and markers noticed that plagiarism occurrences have lowered dramatically, and participants mentioned that it was hard to plagiarize when completing a creative assignment. Finally, the creative assignment was received well by students and it accomplished its goal of reducing plagiarism and allowing students to extend their context into the classroom. We will address concerns raised by students in the next AR cycle.

6 Conclusion

Ultimately, remote emergency teaching and learning provided us with the opportunity to grow as educators, since we had to think creatively about online assessment, as well as the context of the students. An equity-minded approach to assessment in programming education not only created opportunities for our students to bring their context to the classroom, but it also allowed our students to solve problems within their local communities, to foster computational thinking, and to reduce academic dishonesty instances to almost none.

I found it fascinating to be able to create projects that I only even had them as an idea and being given the platform to demonstrate them was an amazing opportunity.

From the comment of the above student, it is recommended to other educators to provide their students with the same opportunity, not only for programming assignments, but we anticipate that the concept of creative assignments is reproducible by educators in other fields.

References

1. Adzima, K.: Examining online cheating in higher education using traditional classroom cheating as a guide. Electr. J. E-Learn. **18**, 476–493 (2020)
2. Barr, V., Stephenson, C.: Bringing computational thinking to K-12: what is Involved and what is the role of the computer science education community? Acm Inroads **2**, 48–54 (2011)
3. Baskerville, R.L.: Investigating information systems with action research. Commun. AIS **2**, 4 (1999)
4. Baskerville, R.L., Wood-Harper, A.T.: A critical perspective on action research as a method for information systems research. J. Inf. Technol. **11**, 235–246 (1996)
5. Berland, M., Wilensky, U.: Comparing virtual and physical robotics environments for supporting complex systems and computational thinking. J. Sci. Educ. Technol. **24**, 628–647 (2015)
6. Creswell, J.W., Clark, V.L.P.: Designing and conducting mixed methods research. (2007)
7. Eaton, S.E.: Academic integrity during COVID-19: reflections from the university of Calgary. Int. Stud. Educ. Adm. **48**, 80–85 (2020)
8. Susman, G.I., Evered, R.D.: An assessment of the scientific merits of action research. Adm. Sci. Q. **23**, 582–603 (1978)
9. Fomunyam, K.G., Teferra, D.: Curriculum responsiveness within the context of decolonisation in South African higher education (2017)
10. Gamage, K.A., Silva, E.K.D., Gunawardhana, N.: Online delivery and assessment during COVID-19: Safeguarding academic integrity. Educ. Sci. **10**(11), 301 (2020)
11. González, M.R.: Computational thinking test: design guidelines and content validation. In: Proceedings of EDULEARN15 conference, pp. 2436–2444. (Year)
12. Gumbo, M.T.: Pedagogical principles in technology education: An indigenous perspective. In: African Indigenous Knowledge and the Sciences, pp. 13–32. Brill Sense (2016)
13. Isomöttönen, V., Tirronen, V.: Flipping and blending—An action research project on improving a functional programming course. ACM Transactions on Computing Education (TOCE) **17**, 1–35 (2016)
14. http://www.iste.org/docs/ct-documents/computational-thinking-operational-definition-flyer.pdf
15. Kazimoglu, C., Kiernan, M., Bacon, L., MacKinnon, L.: Learning programming at the computational thinking level via digital game-play. Procedia Computer Science **9**, 522–531 (2012)
16. Khoza, S.B., Biyela, A.T.: Decolonising technological pedagogical content knowledge of first year mathematics students. Educ. Inf. Technol. **25**(4), 2665–2679 (2019). https://doi.org/10.1007/s10639-019-10084-4
17. Lavonen, J., Autio, O., Meisalo, V.: Creative and collaborative problem solving in technology education: a case study in primary school teacher education. J. Technol. Stud. **30**, 107–115 (2004)

18. Liebenberg, J., Botes, P.: Plagiarism in a GUI programming course. In: 10th Annual International Conference on Computer Science Education: Innovation and Technology (CSEIT), pp. 45–50. Global Science and Technology Forum (GSTF), Bangkok, Thailand (2019)
19. Lye, S.Y., Koh, J.H.L.: Review on teaching and learning of computational thinking through programming: what is next for K-12? Comput. Hum. Behav. **41**, 51–61 (2014)
20. Martin, F., Pirbhai-Illich, F.: Towards decolonising teacher education: criticality, relationality and intercultural understanding. J. Intercult. Stud. **37**, 355–372 (2016)
21. Moten, J., Jr., Fitterer, A., Brazier, E., Leonard, J., Brown, A.: Examining online college cyber cheating methods and prevention measures. Electron. J. E-learn. **11**, 139–146 (2013)
22. Msila, V., Gumbo, M.T.: Africanising the curriculum: indigenous perspectives and theories. African Sun Media (2016)
23. NWU, http://www.nwu.ac.za/gov_man/policy/index.html
24. NWU: North-west university's declaration on the decolonisation of university education. In: University, N.-W. (ed.) (2018)
25. NWU: Teaching, learning assessment policy. In: University, N.-W. (ed.) (2019)
26. https://news.nwu.ac.za/major-data-and-connectivity-boost-nwu-students
27. Reyneke, Y., Shuttleworth, C.C., Visagie, R.G.: Pivot to online in a post-COVID-19 world: critically applying BSCS 5E to enhance plagiarism awareness of accounting students. Acc. Educ. **30**(1), 1–21 (2020)
28. Saurombe, N.: Decolonising higher education curricula in South Africa: factoring in archives through public programming initiatives. Arch. Sci. **18**, 119–141 (2018)
29. Scurr, C.: Reflections on academic integrity during COVID-19. Can. Perspect. Acad. Integrity **3**, 36–38 (2020)
30. Shute, V.J., Sun, C., Asbell-Clarke, J.: Demystifying computational thinking. Educ. Res. Rev. **22**, 142–158 (2017)
31. Susman, G.I., Evered, R.D.: An assessment of the scientific merits of action research. Adm. Sci. Q. **23**(4), 582–603 (1978)
32. Tashakkori, A., Creswell, J.W.: The new era of mixed methods. J. Mixed meth. Res. **1**(1), 3–7 (2007)

Project-Based Learning Guidelines for IT Higher Education

J. T. Janse van Rensburg[✉][iD]

School of Computer Science and Information Systems, North-West University,
Vanderbijlpark Campus, Vanderbijlpark, South Africa
jt.jansevanrensburg@nwu.ac.za

Abstract. It is challenging to adapt higher education at the same pace as the fast-changing nature of the information technology (IT) sector. Creative ways are needed to accommodate the continuous changes in the IT industry, without needing to change the structure of the curriculum. As the IT industry comprises of project teams in project environments, a suitable instructional approach for IT higher education is project-based learning (PBL). A PBL strategy allows for the instruction of current technologies and contributes to skills development that is required in the IT industry. General guidelines and characteristics for PBL are followed across the board for all disciplines on a trial and error basis. Some guidelines may not be directly applicable to IT higher education, while other needed guidelines do not exist because they are domain-specific. The purpose of this paper is to present guidelines for PBL that is relevant to IT higher education. The guidelines are formulated using popular characteristics of PBL from literature and adapting them to the context of IT projects. Additional guidelines are added based on specific requirements of IT projects that are relevant to professional practice. A literature review matrix is presented as evidence of sources where the guidelines were adapted from. The guidelines can be used by educators who find it challenging to assess whether they are implementing a suitable PBL approach. Additionally, following the guidelines may contribute towards bridging the skills gap between IT higher education and the IT industry.

Keywords: Project-based learning · Information technology · IT higher education · Guidelines

1 Introduction

The goal of this article is to formulate guidelines for effectively using project-based learning, specifically in the information technology (IT) higher education sector. The need for the guidelines is emphasised in research that indicates how challenging it can be for facilitators to assess whether they are implementing an acceptable approach to project-based learning (PBL) [1,2]. Without clear guidance of what a project-based learning strategy should resemble, it is challenging for facilitators to evaluate the quality of their PBL approach, or to understand where improvement is needed [3]. More so, "education policymakers are increasingly demanding evidence to guide decisions about whether to adopt an educational reform or instructional innovation" [3].

Project-based learning emphasises deeper learning to develop skills relevant to university life and future careers [4,5]. The goals of a PBL approach align well with the development of higher-level thinking skills and interpersonal skills [3]. Specifically in IT higher education, project-based learning shows evidence of developing relevant IT skills, teamwork, communication, time management, and collaborating with diverse individuals [6]. Project-based learning is a student-centered approach that targets advanced skills such as analysis, evaluation, and synthesis, but can also be supported by teacher-centered approaches to communicate required knowledge, and to demonstrate specific skills [7].

In this article, a general overview of the background of project-based learning is provided. A discussion on how project-based learning has been applied in IT higher education, and similar fields, follow. This leads into a discussion on the importance of reflective practice in project-based learning. Thereafter, a set of coherent guidelines for project-based learning in IT higher education is presented which were formulated from PBL characteristics in literature, as well as core IT project requirements. The set of guidelines are given in response to the literature that indicates how challenging it can be to assess whether an acceptable approach to PBL is being implemented without guidance.

2 PBL Overview

"Project-based learning is a model that organises learning around projects" [1]. The first foundational introduction of project-based learning was by William Heard Kilpatrick [8], which referred to it as the 'project method', and was inspired by the philosophies of John Dewey [9]. Kilpatrick's idea of the project method centered on "an activity undertaken by students that really interested them" [10]. The activity was typically in the form of a project, with four distinctive student-centered phases: purposing, planning, executing, and judging [11]. Thomas and Kilpatrick both stressed that intrinsic motivation, deeper learning, and student-centered inquiry were central benefits of project-based instruction [12]. Project-based learning (PBL) is based on three principles: learning takes place for a specific context, active participation in the learning process is expected, and learning goals are reached though the sharing of knowledge in social interactions [13]. PBL is an instructional methodology that facilitates student learning through student-centered projects [14], and is often cited as a preferred method to traditional teaching strategies. Comparative studies of PBL and traditional instruction offer insight into the significance of project-based learning as preferred method to master content and relevant skills [11]. However, traditional instruction is perceived to have less management responsibilities than project-based instruction has. Other authors have made the argument that PBL is no more demanding than traditional instruction, can be implemented with limited resources, and within the same timeframe [15]. Project-based learning is seen as superior in some aspects to traditional summative assessment, as it is designed to deliver outcomes not adequately evaluated through traditional methods [14,16]. Project-based learning often uses performance-based assessments, which provides a platform for evaluating 21st century competencies such as critical thinking, reflection, and problem solving–which are difficult to measure [17]. Project-based learning exposes students to a variety of workplace skills such as communication, teamwork, self-management, flexibility, and collaboration [3,18,19]. Through PBL, students learn to be independent by setting goals, planning and organising project requirements, developing collaboration skills, and learning at their own level [20]. Project-based learning can be an effective learning tool for lower achieving students, "students with special needs, low-income students, and students with limited English proficiency" [3]. PBL has been used in many different environments, ranging from primary school, to secondary school, to higher education [12]. PBL is inquiry-based instruction where authentic, real-world problems drive the learning context and lead to meaningful experiences [15,21]. The value of project-based learning lies in the construction of a concrete artefact or end product [12,22,23]. The skills trained through artefact creation assists in self-regulation and taking ownership of the learning process [24], with the aim of students becoming self-directed whilst applying higher order thinking skills [11]. PBL delivers increased levels of student engagement, problem-solving skills development, transferability of skills, and an improved interest in the subject matter [1,25]. Successful project-based learning relies on the facilitator's ability to provide motivation, guidance, and scaffolding appropriate to the context of the students [12]. Facilitators who support the development of skills, instead of focusing on standard assessment achievements, can more easily implement and align with a PBL curriculum, however, developing a coherent PBL curriculum is one of the most difficult aspects of project-based learning [3]. A literature review of interdisciplinary project-based learning as an instructional method for a variety of undergraduate engineering and computer science degrees highlighted the need for collaboration with industry for skills development, embedding employability skills in curriculum planning, and developing effective methods for assessing employability skills [26]. Project-based learning has been widely adopted to develop employability skills in universities, indicating that discipline content need not be sacrificed for the development of employability skills when project-based learning is used effectively.

3 Applications of PBL in IT Higher Education

The use of PBL is popular in engineering, computer science, and information technology degrees due to the expected artefact creation that is typical of these project-based environments. One example is a model for project-based learning that was implemented in two engineering subjects [27]. Seventy students provided feedback on the impact of the PBL strategy taken. The study concluded that team-based PBL is a suitable pedagogy for improving student engagement, and supporting skills such as problem-solving, report writing, and critical thinking. Project-based learning is a practical approach to skills development in any field and can be used to evaluate the skill levels of students in a collaborative environment that simulates an industry setting. A PBL approach in computer science provides an environment to practice and develop workplace skills that are challenging to simulate in traditional lecture settings [28]. PBL promotes professional skills, discipline and time management skills, in a similar manner as the professional world would demand. For IT projects, the use of agile methodologies to manage PBL teams in computer science education is recommended [29].

Project-based learning does have challenges in its implementation, specifically the manner in which tasks are divided amongst teams in a project, as there is a perceived dissatisfaction in teams where two or more women are grouped with men [30]. Male students tend to take on the role of programmer, while female students are tasked with the role of documenter. Men are more likely to gain technical experience through PBL activities, while women more often build on project management and communication skills [30]. While the acquisition of any new skills is a step in the right direction, the structure of PBL activities should ensure that technically skilled students are also produced from the historically underrepresented groups in computer science and engineering. A gap between academic education and the role of the university was also noted where excellent programmers fail a degree due to poor perceived academic performance, while students who struggled with coding passes a degree based on good academic performance [29]. This issue emphasises the need for more practical based curriculum in IT degrees, considering that the field of information technology is constantly evolving. Research on the benefits and challenges of project-based learning in IT higher education needs further development.

The use of projects in IT higher education is expected due to the fact that the IT industry is comprised of project environments [31]. This suggests that a project-based learning environment has the potential to not only train technical IT skills, but also contribute to a robust understanding of entrepreneurship, data analysis, computational thinking, critical thinking, creativity, and other 21st century skills. Project-based learning is a suitable pedagogy that can address the IT skills gap experienced worldwide [32–35], as it supports the IT graduate's development through real-life experiences [31]. For best practice, a project-based learning environment should encourage continuous reflective practice for professional development [3,36,37].

4 Reflective Practice in PBL

At its core, project-based learning is a reflective practice approach to solving a project-based problem. In PBL, active engagement and reflection contributes to experiential learning as students need to collaborate on, and contribute to, the outcome of the project [22]. It is difficult to assess the reflection that takes place on learning [6], but research continuously places emphasis on the importance of reflection, self-evaluation, and peer assessment in project-based learning [20,37–39]. It is crucial to make time for "students to reflect deeply on the work they are doing and how it relates to larger concepts specified in the learning goal" [39].

Students' conceptual knowledge is developed through a systematic process of documenting the learning process, and reflecting on the experience [40]. It is important that both the learner and the facilitator take time to reflect. "Throughout a project, students - and the teacher - should reflect on what they're learning, how they're learning, and why they're learning" [37]. Reflection can take place through methods such as autoethnography and reflective journaling [41].

Students should be encouraged to think about what they are doing while they are doing it, and after they have done it, to improve the manner in which they will address similar problems in future [42]. This approach of dual reflection is referred to as reflective practice. Reflective practice is a professional learning and development strategy focused on improving practice. It is "a dialog of thinking and doing through which I become more skilful" [42]. Even when professional practitioners engage in problem solving, they do so through reflection. Research on the implementation of project-based learning strongly suggests that the reliability of a PBL strategy centres on its contribution to professional development [3]. Reflective practice is central to professional development, and is considered as a method to transform implicit knowledge into explicit knowledge through experience [43]. For this reason, reflective practice is considered central to the success of a project-based learning strategy, towards cultivating professional development in students.

5 Guidelines for PBL in IT Higher Education

Thomas [1] made an observation two decades ago that is still problematic today–"there continues to be a lack of consensus on what constitutes PBL" [3]. Condliffe and colleagues [3] recommend aspects to be cautious of when constructing design principles for PBL:

- Design principles should be measurable,
- Design principles should address both content and assessment,
- Design principles should be informed by practice,
- Research on design principles should investigate adaptation.

Following the recommendations made above, Table 1 provides a coherent set of guidelines that can be used for effective project-based learning in IT higher education. The information is presented as a literature matrix where well-cited PBL sources were consulted. Information technology sources were included to improve rigour for IT projects. It is important to note that there are limited PBL guidelines specifically for IT education in the literature, where most only exist from research that applies generic PBL principles. As such, the guidelines presented below were formulated using the general characteristics of what constitutes good PBL, and were adapted to the scope of information technology projects. Additional guidelines that are specific to IT projects were added for value. The aim of the guidelines, if implemented properly, is to ensure greater alignment between IT higher education and the IT industry. This in return will develop the necessary skills that make IT students more employable.

Table 1. PBL guidelines for IT higher education.

PBL guidelines for IT projects in higher education	44	45	46	3	12	36	2	20	37	22	23	47	38	14	1	48	49
The primary objective of a PBL environment for IT students is to focus on improving specific 21st century competencies and IT skills	x	x	x	x	x	x	x	x	x	x	x	x	x	x	x	x	x
Project-based learning is central to the IT curriculum, and is not viewed as a separate activity at the end of a module	x	x	x	x		x		x				x		x	x	x	x
Every IT project has specific educational goals as set out in the curriculum	x	x	x	x					x		x				x	x	x
An IT project in a PBL environment is designed to encourage reflective practice in both the student and the facilitator			x	x	x	x	x	x	x	x	x	x	x	x	x	x	x
For an IT PBL environment to be successful, support is needed from colleagues and faculty management	x	x		x	x							x		x			x
The scope of an IT project should be significant, realistic and industry related	x	x	x	x	x	x	x	x	x	x	x	x	x	x	x	x	x
IT students should be allowed to have a voice and choice appropriate to the context of the project	x		x	x	x		x	x	x	x					x	x	x
IT projects should result in IT artefacts	x	x	x	x		x	x	x		x	x	x		x	x	x	x
The use of appropriate technology in IT projects is expected and not a nice to have	x	x	x	x	x	x	x	x	x	x			x	x	x	x	
The project scope should be well defined with enough time allocated for in-depth inquiry	x	x		x	x	x	x	x	x	x	x		x	x	x	x	x
An IT project should start with a driving question that makes the student think about the project over a period of time before actually starting with the creation of an artefact				x	x		x	x	x	x	x		x	x	x	x	x
Intrinsic motivation for IT students and the need to know are reinforced with tools such as guest lectures from industry or building real solutions for actual projects				x	x	x	x	x	x	x			x	x	x	x	x
Facilitator guidance should be appropriately managed with scaffolding sources that further improve the need to know of the student	x			x	x	x	x	x	x		x	x	x	x	x	x	x
Teamwork is emphasised to simulate working environments in the IT industry	x	x	x	x	x	x	x	x		x	x	x	x	x	x	x	x
Learning by doing is emphasised through active participation throughout the IT project	x	x		x			x					x	x	x	x	x	x
Time management is crucial in IT projects	x	x	x	x	x		x	x	x	x			x	x	x	x	
Use appropriate agile methodologies to manage IT teams		x	x														
Encourage the use of source control software to manage and track code changes, and to determine team member contribution	x																
User stories will guide the creation of simplified software requirements in an IT project		x	x														
Assessment practices for IT projects should include critique, revision, and rubrics	x		x	x	x	x	x	x	x	x		x	x	x	x	x	x
Motivation to deliver quality IT projects is improved when external audiences such as community or industry members are involved and act as evaluators of the artefact		x		x	x	x		x	x	x	x	x		x	x		x

While compiling the guidelines, it was evident that they could be grouped under different phases of a PBL activity. As such, a discussion of the guidelines in context of IT higher education follows, and is structured according to a suggested PBL activity process flow (Fig. 1).

Fig. 1. Process flow for PBL guidelines

5.1 Reasoning and Reflection

Guidelines that form part of the reasoning and reflection criteria provide context to the objectives of adopting a project-based learning strategy.

The primary objective of a PBL environment for IT students is to focus on improving specic 21st century competencies and IT skills - Before planning a project-based learning activity, ask yourself, "which skills do I hope will be developed through this PBL activity?" The nature of a project-based learning environment lends itself to skills development, but it is up to the facilitator to determine which skills need to be developed. As a PBL environment resembles an IT industry environment, specific skills development activities can be incorporated with relative ease. For example, weekly demonstrations to train soft skills, and specific development platforms to train technical skills.

Project-based learning is central to the IT curriculum, and is not viewed as a separate activity at the end of a module - When planning to use a project-based learning approach, you need to begin with the end in mind. The project is the central activity in the module, and all other activities are planned in such a way that it supports the development and improved outcome of the project. Educators tend to go about their semester in the traditional manner, and then add a project that covers some of the aspects that were discussed during the semester. This is not the correct approach to PBL (as indicated by Thomas). A successful PBL strategy is goal-driven, and is built around the project.

Every IT project has specific educational goals as set out in the curriculum - All modules that form part of a degree have module outcomes. Furthermore, all study units that form part of a module have study unit outcomes. When formulating your project-based learning activity, you should take cognisance of the learning outcomes that need to be met. This is an important element that forms part of the administrative documentation of the institution, where a coherent strategy must be documented demonstrating how module objectives were met.

An IT project in a PBL environment is designed to encourage reflective practice in both the student and the facilitator - A project-based learning activity is formulated through reflective practice, and encourages reflective practice. PBL activities for IT projects aims to improve the professional development of the student, by guiding the student to become a reflective practitioner. The PBL environment also improves the reflective practice of the facilitator, who needs to adapt her problem-solving skills in the context of each individual project.

5.2 Requirements

Guidelines that form part of the requirements criteria provide context to aspects that need to be considered before planning a project-based learning activity.

For an IT PBL environment to be successful, support is needed from colleagues and faculty management - Faculty support is important when specific tools and platforms are required. You may also need other consumable resources such as brown paper and stationery for creative planning sessions. Support may also be needed when the facilitator wants to attend workshops or industry training sessions to improve her own skills related to facilitating the project-based learning activity. Support from colleagues may be required in the form of evaluation assistance, as assessing PBL artefacts can be time-consuming.

The scope of an IT project should be signicant, realistic and industry-related - It is recommended to use industry project scopes as part of a project-based learning environment for IT students. Using authentic project briefs will encourage the commitment students have towards completing the project. Community projects are also ideal for PBL activities if there is a need in the community, because the students then feel that they are contributing to a real solution. If it is not possible to obtain a project scope from industry/the community, do market research on industry-relevant topics that are currently in the headlines, preferably using wicked problems that will develop the critical thinking and problem solving skills of the student.

IT students should be allowed to have a voice and choice appropriate to the context of the project - Entry-level and exit-level IT students are motivated by different types of projects. For entry-level students, consider using a voting platform where students can suggest, and then vote for, the project scope that they would like as a topic. The facilitator should shortlist the topics based on their suitability and relevance to most/all students. For exit-level students, a multifaceted project scope can be used that explains the industry problem in context of different role players, where an administrator might need a desktop application, a client might need a web application, and a driver might need a mobile application. The student can then choose which role player's problem s/he wants to solve.

IT projects should result in IT artefacts - A demonstrable artefact is expected at the end of an IT project-based learning activity. These include, but are not limited to, software documentation, systems, applications, and research reports.

The use of appropriate technology in IT projects is expected and not a nice to have - It is important to integrate authentic software and tools as part of your IT project-based learning activity. From the market research you've conducted to determine which tools and software are appropriate, you should be able to make a list of requirements for faculty management. For IT degrees, faculties usually have specific budgets available for upgrade and maintenance of the IT infrastructure. Some technology stacks offer educational licenses, such as Visual Studio. There are also open source platforms that are widely used and that can be implemented, such as Python.

5.3 Design and Scheduling

Guidelines that form part of the design and scheduling criteria provide context to aspects that should be implemented while planning a project-based learning activity.

An IT project should start with a driving question that makes the student think about the project over a period of time before actually starting with the creation of an artefact - The driving question is a neutral open-ended question that will lead students to investigate processes or procedures that might be useful for the PBL activity. The driving question is given at the start of a module, or before the project commences, to focus the direction of the investigation. An example of a driving question is, "where can I host a website for free that will include free database storage of at least X megabytes per month?" Students will start researching possible hosting platforms without needing to know what the project is about. The open-ended question will also keep them thinking about the project, holding interest and motivation for when the project scope is released.

The project scope should be well defined with enough time allocated for in-depth inquiry - There should be no ambiguities in the project scope. This does not necessarily mean that you need to provide a bullet list of instructions. An effective manner of structuring briefs for IT projects is in narrative format. Write a problem statement from the views of the respective role players in the project. The time allocated to completing the project should be fair, and can only really be assessed through active involvement of the facilitator. IT projects are generally scheduled between 4–6 weeks (depending on the scope), with the assumption that the students have dedicated development time during every week where facilitator guidance is available.

5.4 Guidance and Scaffolding

Guidelines that form part of the guidance and scaffolding criteria provide context to the manner in which the facilitator should provide support and direction in the project-based learning activity.

Intrinsic motivation for IT students and the need to know are reenforced with tools such as guest lectures from industry or building real solutions for actual projects - For increased motivation, arrange an industry guest lecture that accompanies the industry project scope. Students are inspired by practitioners, and will reflect deeply on the experience. Real projects for real clients also improve motivation to complete the project, especially community projects that are completed with the aim of alleviating a real problem. For example, a local animal shelter that requires a free website where they can upload pictures of pets that are up for adoption.

Facilitator guidance should be appropriately managed with scaffolding sources that further improve the need to know of the student - When structuring project scopes, make sure to add web links to resources that will be useful to complete the project. Also provide links to other similar projects, or similar problems experienced, to motivate the student's search for solutions. As part of the planning for a PBL activity, the supporting activities can also be scaffolded in such a way that an important piece of information is revealed every week to maintain the interest of the students.

5.5 Monitoring and Management

Guidelines in the monitoring and management criteria provide guidance for important aspects that should form part of IT projects towards the professional development and employability of the IT student.

Teamwork is emphasised to simulate working environments in the IT industry - Where possible, team projects should be incorporated when using project-based learning. The IT industry comprises of IT teams that create IT projects. Students need to work in teams where they are unfamiliar with their team members. Working with new people will develop teamwork and time management skills, more so than working in teams with friends. A suggested number of members per team is 4–taking on roles such as the business analyst, back-end developer, front-end developer, and database administrator.

Learning by doing is emphasised through active participation throughout the IT project - Project-based learning aims to encourage student reflection. There is a transition from 'knowing through making' the artefact, to 'learning by doing' what is best practice for developing artefacts. Students that do not actively participate in the project will miss out on the chance to learn from peers and mentors in a controlled environment.

Time management is crucial in IT projects - One of the biggest challenges for IT graduates is to adjust to the demanding deadlines of the IT industry. Training time management is essential in an IT PBL activity, and can be coached by setting time constraints on different activities that form part of the project, weekly feedback sessions, and time boxing demonstrations so that students learn how to highlight the most important items in their projects first.

Use appropriate agile methodologies to manage IT teams - IT companies use agile development methodologies, such as DevOps, Extreme Programming, Scrum, and Kanban, to manage their workflow. Educating students on the methodologies that are available, and providing training in at least one method is recommended. For example, Scrum is a useful methodology that demonstrates how to plan backlog items, sprints, and daily stand-up meetings. The approach can be supported by using a visual work-pull system such as a Kanban board that tracks the progress and contribution of all team members. Brown paper sheets and sticky notes work well for a tactile display, or alternatively an online platform such as Trello can be used.

User stories will guide the creation of simplified software requirements in an IT project - Creating user stories is a standard approach to defining software requirements for a system. A user story typically describes the type of user, what the user wants to do, and why the user wants to do it. From a user story, functional and non-functional requirements of the system can be specified. To mimic professional practice, it is important to emphasise the development of core functional requirements first. For example, if the client wants an online booking system for their hotel, the core functional requirement of the system would be the ability to make a booking. Items such as e.g. registration and login features are secondary, and should be focused on after the booking item is functional. Discourage the development of unnecessary 'nice to haves' that were not part of the project brief, or postpone development thereof until all core functional requirements have been implemented first.

Encourage the use of source control software to manage and track code changes, and to determine team member contribution - A standard requirement in IT practice is the use of source control. Source control, or version control, is a software program that manages and tracks code changes, providing a history of code that was developed, logged, and merged. The benefits of source control include access to previous versions of a project that was in workable condition, and shared access in teams to work on the same code. Examples of source control platforms are Git, Azure DevOps, and BitBucket. In IT student projects, the extended benefit of implementing source control, is the ability of the facilitator to see how long the team has been working on the project, and who has been working on the project. The software keeps track of the dates when code was committed, and by who the code was committed. Another benefit is that students cannot use the excuse that "my project was working just this morning", because your response will be "then run your previous code commit that was working on your source control platform".

5.6 Assessment

Guidelines that form part of the assessment criteria provide recommended evaluation practices for IT projects.

Assessment practices for IT projects should include critique, revision and rubrics - Project assessment during development should include 5–10 min stand-ups where a student/team highlights what has been completed so far, items that were struggled with, and what needs to happen next. This activity develops their verbal communication and presentation skills. It is recommended that every project brief should have an accompanying project rubric. A detailed rubric provides the students with a checklist of items that need to be completed, and eliminates the possibility of inconsistent marking as all students/teams are scored using the same criteria.

Motivation to deliver quality IT projects is improved when external audiences such as community or industry members are involved and act as evaluators of the artefact - It is recommended that public audiences take part in the evaluation of the final artefact. Students will be motivated to produce quality artefacts when they know that members from the industry will be evaluating their work, and in turn, this will provide industry with an opportunity to scout for talent. When using community projects, members from the community can also be invited to evaluate the artefact. Careful consideration should be given to how the marks are awarded when using novice or unknowledgeable judges from the community. If you are unable to arrange public audiences, ask colleagues for assistance with evaluating the projects. An unfamiliar face will encourage professional demonstrations.

6 Conclusion

This paper presented a set of guidelines for using project-based learning effectively in IT higher education. The guidelines were formulated from existing characteristics of PBL in literature, and were supplemented with guidelines that are specifically relevant for IT projects. The novelty of the paper rests specifically on the guidelines that were formulated as part of the monitoring and management of IT PBL activities. These guidelines are specific to the IT sector, and emphasises aspects that should be part of IT PBL activities to mimic professional practice. Figure 2 depicts a high-level overview of the PBL guidelines presented in this article.

Fig. 2. Guidelines for PBL in IT higher education

The need for the guidelines was emphasised in literature, where it was highlighted that educators find it challenging to assess whether a suitable PBL approach is being implemented. After reflecting on the possible impact of the guidelines, it was noted that universities might be placing too much emphasis on theoretical work at the disadvantage of practical experience expected in an IT degree. Project-based learning is a solution towards including more practical content in an IT degree, as well as preparing students for industry. Following the PBL guidelines proposed in this research may contribute towards bridging the gap between IT theory (higher education) and IT practice (industry).

References

1. Thomas, J.W.: A Review of Research on Project-based Learning, Autodesk Foundation, San Rafael (2000)
2. Tamim, S.R., Grant, M.M.: Definitions and uses: case study of teachers implementing project-based learning. Interdiscipl. J. Probl.-Based Learn. **7**(2), 3 (2013)
3. Condliffe, B., et al.: Project-Based Learning: A Literature Review. MDRC, New York (2017)
4. Huberman, M., et al.: The Shape of Deeper Learning: Strategies, Structures, and Cultures in Deeper Learning Network High Schools, Institutes for Research, Washington, DC (2014)
5. Scardamalia, M., et al.: New assessments and environments for knowledge building new assessments and environments for knowledge building. In: Griffin, P., McGaw, B., Care, E. (eds.) Assessment and Teaching of 21st Century Skills. Springer, Dordrecht (2012). https://doi.org/10.1007/978-94-007-2324-5_5
6. Whatley, J.: Evaluation of a team project based learning module for developing employability skills. Iss. Inform. Sci. Inf. Technol. **2012**(9), 75–92 (2012)
7. Veletsianos, G., et al.: Design principles for thriving in our digital world: a high school computer science course. J. Educ. Comput. Res. **54**(4), 443–461 (2016)
8. Kilpatrick, W.H.: The project method. Teacher's college. Record **19**, 319–335 (1918)
9. Dewey, J.: Democracy and Education: An Introduction to the Philosophy of Education, p. 434. Macmillan, New York (1916)
10. Ravitch, D.: Left Back: A Century of Failed School Reforms. Simon and Schuster, New York (2000)
11. Holm, M.: Project based instruction: a review of the literature on effectiveness in prekindergarten. River Acad. J. **7**(2), 1–13 (2011)
12. Kokotsaki, D., Menzies, V., Wiggins, A.: Project-based learning: a review of the literature. Improv. School **19**(3), 267–277 (2016)
13. Cocco, S.: Student Leadership Development: the Contribution of Project-based Learning. Royal Roads University, Victoria, BC (2006)
14. Mergendoller, J.R., Thomas, J.W.: Managing project based learning: Principles from the field. Buck Institute for Education (2000). http://www.bie.org
15. Al-Balushi, S.M., Al-Aamri, S.S.: The effect of environmental science projects on students' environmental knowledge and science attitudes. Int. Res. Geogr. Environ. Educ. **23**(3), 213–227 (2014)
16. Hertzog, N.B.: Transporting pedagogy: implementing the project approach in two first-grade classrooms. J. Adv. Acad. **18**(4), 530–564 (2007)

17. Pellegrino, J.W., Hilton, M.L.: Education for Life and Work: Developing Transferable Knowledge and Skills in the 21st Century. National Academies Press, Washington, DC (2012)
18. Musa, F., et al.: Project-based learning (PjBL): inculcating soft skills in 21st century workplace. Procedia-Soc. Behav. Sci. **59**, 565–573 (2012)
19. Musa, F., et al.: Project-based learning: promoting meaningful language learning for workplace skills. Procedia-Soc. Behav. Sci. **18**, 187–195 (2011)
20. Bell, S.: Project-based learning for the 21st century: skills for the future. The Clear. House **83**(2), 39–43 (2010)
21. Wurdinger, S., et al.: A qualitative study using project-based learning in a mainstream middle school. Impro. Schools **10**(2), 150–161 (2007)
22. Helle, L., Tynjälä, P., Olkinuora, E.: Project-based learning in post-secondary education - theory, practice and rubber sling shots. Higher Educ. **51**(2), 287–314 (2006)
23. Krajcik, J.S., Blumenfeld, P.C.: Project-based learning (2006)
24. Ertmer, P.A., Simons, K.D.: Scaffolding teachers' efforts to implement problem-based learning. Int. J. Learn. **12**(4), 319–328 (2005)
25. Barron, B.J.S., et al.: Doing with understanding: lessons from research on problem and project-based learning. J. Learn. Sci. **7**(3/4), 271–311 (1998)
26. Hart, J.L.: Interdisciplinary project-based learning as a means of developing employability skills in undergraduate science degree programs. J. Teach. Learn. Grad. Employ. **10**(2), 50–66 (2019)
27. Viswambaran, V.K., Shafeek, S.: Project based learning (PBL) approach for improving the student engagement in vocational education : an investigation on students' learning experiences & achievements. In: 2019 Advances in Science and Engineering Technology International Conferences (ASET). Dubai, United Arab Emirates (2019)
28. McManus, J.W., Costello, P.J.: Project based learning in computer science: a student and research advisor's perspective. J. Comput. Sci. Coll. **34**(3), 38–46 (2019)
29. Renz, J., Meinel, C.: The "Bachelor Project": project based computer science education. In: 2019 IEEE Global Engineering Education Conference (EDUCON). IEEE (2019)
30. Fowler, R.R., Su, M.P.: Gendered risks of team-based learning: a model of inequitable task allocation in project-based learning. IEEE Trans. Educ. **61**(4), 312–318 (2018)
31. Sindre, G., et al.: Project-based learning in IT education: definitions and qualities. UNIPED **2018**, 27 (2019)
32. CompTIA. Assessing the IT skills gap. 2017 5 November 2019. https://www.comptia.org/content/research/assessing-the-it-skills-gap
33. Cobb, S.: Mind this gap: criminal hacking and the global cybersecurity skills shortage, a critical analysis. In: Virus Bulletin Conference (2016)
34. Van Broekhuizen, H.: Graduate unemployment and higher education institutions in South Africa. In: Bureau for Economic Research and Stellenbosch Economic Working Paper 08, vol. 16 (2016)
35. Schofield, A., Dwolatzky, B.: 2019 JCSE-IITPSA ICT skills survey, 5 November 2019. The Tenth Edition. https://www.iitpsa.org.za/wp-content/uploads/2019/09/2019-JCSE-IITPSA-ICT-Skills-Survey-v1.pdf
36. Lee, J.S., et al.: Taking a leap of faith: redefining teaching and learning in higher education through project-based learning. Interdiscipl. J. Probl.-Based Learn. **8**(2), 2 (2014)
37. Larmer, J., Mergendoller, J.R.: Essentials for project-based learning. Educ. Leadersh. **68**(1), 34–37 (2010)

38. Grant, M.M.: Getting a grip on project-based learning: theory, cases and recommendations. Meridian **5**(1), 83 (2002)
39. Darling-Hammond, L.: Teaching and learning for understanding. In: DarlingHammond, L., et al. (eds.) Powerful Learning: What We Know About Teaching for Understanding, pp. 1–8. JosseyBass, San Francisco (2008)
40. Barak, M.: From "doing" to "doing with learning": reflection on an effort to promote self-regulated learning in technological projects in high school. Eur. J. Eng. Educ. **37**(1), 105–116 (2012)
41. Krajcik, J.S., et al.: A collaborative model for helping middle grade science teachers learn project-based instruction. Element. Sch. J. **94**(5), 483–497 (1994)
42. Schön, D.A.: The Reflective Practitioner: How Professionals Think in Action. Basic books, London (1983)
43. Finlay, L.: Reecting on 'Reective practice'. Practised-based professional learning center, vol. 29, p. 2014 (2008)
44. Milentijevic, I., Ciric, V., Vojinovic, O.: Version control in project-based learning. Comput. Educ. **50**(4), 1331–1338 (2008)
45. Mahnič, V.: The capstone course as a means for teaching agile software development through project-based learning. World Trans. Eng. Technol. Educ. **13**(3), 225–230 (2015)
46. Kastl, P., Romeike, R.: Agile projects to foster cooperative learning in heterogeneous classes. In: 2018 IEEE Global Engineering Education Conference (EDUCON). IEEE (2018)
47. Doppelt, Y.: Implementation and assessment of project-based learning in a flexible environment. Int. J. Technol. Des. Educ. **13**(3), 255–272 (2003)
48. Moursund, D.G.: Project-based Learning Using Information Technology. International Society for Technology in Education, Eugene (1999)
49. Barron, B.J., et al.: Doing with understanding: lessons from research on problem- and project-based learning. J. Learn. Sci. **7**(3–4), 271–311 (1998)

Teaching Methods and Strategies

Mapping Computational Thinking Skills to the South African Secondary School Mathematics Curriculum

Karen Bradshaw[1,2] and Shannon Milne[1,2]

[1] Department of Computer Science, Rhodes University, Makhanda, South Africa
[2] Department of Information Systems, Rhodes University, Makhanda, South Africa
k.bradshaw@ru.ac.za

Abstract. Computational thinking (CT) is gaining recognition as an important skill for learners in both Computer Science (CS) and several other disciplines, including mathematics. In addition, researchers have shown that there is a direct correlation between poor mathematical skills and the high attrition rate of CS undergraduates. This research investigates the use of nine core CT skills in the South African Grades 10–12 Mathematics curriculum by mapping these skills to the objectives given in each of the topics in the curriculum. The artefact developed shows that all the identified CT skills are used in the curriculum. With the use of this mapping, future research on interventions to develop these skills through mathematics at secondary school, should produce school leavers with better mathematical and problem solving abilities, which in turn, might contribute to better success rates in CS university courses.

Keywords: Computational thinking · Secondary school mathematics curriculum · Computer science

1 Introduction

Computational thinking (CT) as defined by Wing is *"the thought processes involved in formulating problems and their solutions so that the solutions are represented in a form that can be effectively carried out by an information-processing agent"* [13]. Wing further elaborates on how she believes that CT involves all critical thinking skills, including problem solving and designing systems that draw on concepts from mathematics and engineering [12].

As cited by Mwakapenda, the South African Department of Education (DoE) in its National Curriculum Statement (NCS) in 2003 defines the mathematics discipline as follows: *"Mathematics enables creative and logical reasoning about problems in the physical and social world and in the context of mathematics itself. It is a distinctly human activity practised by all cultures ... Mathematics is based on observing patterns; with rigorous logical thinking, this leads to theories of abstract relations. Mathematical problem solving enables us to understand the world and make use of that understanding in our daily lives."* [7, p. 190].

In order to link this definition of mathematics to CT concepts, problem solving (PS) needs to be defined clearly. Voskoglou and Buckley [10] outline how problem solving is used within the context of CT. They describe PS as an activity that makes use of cognitive and/or physical means to solve a problem in order to develop a better idea of the world. PS forms the essence of mathematics, which corresponds with the view of mathematics held by the DoE. PS also forms part of the essence of Wing's definition of CT [12]. In their conclusion, Voskoglou and Buckley [10] give their view on how mathematics should be taught in the future, namely, that there should be a focus on teaching not only mathematical concepts and techniques of computing, but also on PS in order to satisfy aims such as creativity, systematisation, communication, argumentation, and seeing how mathematics can be applied in the real world.

Not only do mathematical concepts underpin many computing and programming concepts, but without the skills developed in studying mathematics, it would be a challenge for potential computer science (CS) students to make sense of abstract language, work with algorithms, accurately model and solve real-life problems and be able to self analyse their work. It has been shown in the literature that poor mathematical skills and problem solving abilities are one of the main reasons for the high attrition rates in CS courses [2]. Given that entry to most CS Bachelor degrees in South Africa requires at least matriculation mathematics, it would be beneficial to those learners who were considering taking CS at university, that not only their knowledge of mathematical concepts, but also their problem solving and other skills associated with studying mathematics, were developed to the fullest during their school years.

To understand exactly what these necessary skills are and where they occur in the South African secondary school mathematics curriculum, this study set out to map core CT skills identified in the literature to the existing mathematics curriculum. Once the skills appropriate to each topic in the mathematics curriculum are known, future research can focus on designing exercises/study aids that can develop and/or enhance such skills to assist learners that struggle with comprehending the core mathematics content and in so doing, develop undergraduates that have the requisite skills to succeed in CS courses.

In the rest of this paper, we discuss related work in the application of CT in various disciplines (Sect. 2), introduce the core set of CT skills used in the study (Sect. 3), present the mapping of these skills to some of the topics in the mathematics curriculum (Sect. 4), and finally in Sect. 5, state our conclusions with suggestions for future work.

2 Computational Thinking in the Context of STEM

Computational thinking draws from concepts of CS such as, algorithms, problem decomposition and simulation to solve problems that often transcend the CS discipline [4]. On the other hand, Howland et al. [3] postulate that as CS students learn programming languages they pick up high-level skills that are broadly applicable outside of CS, including the ability to define clear, specific and unambiguous instructions for carrying out a process; the ability to design a system made up of distinct components with clearly defined areas of responsibility; the ability to design a system made up of components that deliberately reveal certain information and operations related to their purpose, and hide everything else; and the understanding that a complex system of behaviours can arise from simple interactions. Thus, there is a direct reciprocal relationship between CT and CS, in that enhanced CT skills improve CS understanding, while CS programming knowledge develops specific problem solving skills associated with CT. Exploring this relationship further, several studies have shown that CT and STEM[1] subjects generally share a similar reciprocal relationship [14].

The ideas discussed above are in line with Wing's sentiments [12] that CT is a fundamental skill for solving complex problems encountered in everyday life not just in the field of CS. Wing further postulates that CT can enhance skills that are currently required in school curricula. These skills include finding solutions to given problems and designing intelligent solutions using different levels of abstraction and decomposition, algorithms and mathematical concepts while understanding and determining the impact of the solutions chosen for their efficiency, economic and social impacts [12].

Based on the belief that embedding CT in a curriculum requires a practical approach grounded in an operational definition, Barr and Stephenson set out to answer questions such as *"What would computational thinking look like in the classroom? What are the skills that students would demonstrate? What would a teacher need in order to put computational thinking into practice? What are teachers already doing that could be modified and extended?"* [1]. By answering these questions the authors proposed nine key CT concepts and capabilities (as explained in Sect. 3) and provided linkages of the identified skills to other disciplines including mathematics.

A different approach was used in [8] to investigate CT education in K-12 German secondary schools. The methodology used was based on the analysis of secondary data, namely the curricula frameworks for CS and Physics in four federal German states. The authors also compared CT education in Germany with that at the international level by considering the ACM model curriculum for K-12 computer science. The state curricula were analysed and coded based on whether the vocabulary in the description of competencies or learning objectives in the curriculum frameworks matched specific CT terms and components. The results were presented in the form of a table showing whether the curriculum of each state contained the nine core computational concepts as defined by Barr and Stephenson [1].

Weintrop et al. [11] stated that due to the increased presence of computation in mathematics and scientific contexts there is an urgency to define CT and provide a theoretical grounding for how it should be applied in school science and mathematics classrooms. To do this, they proposed a taxonomy consisting of four main categories: data practices, modeling and simulation practices,

---

[1] STEM is an acronym for four closely connected areas of study, namely, science, technology, engineering and mathematics.

computational problem solving practices, and systems thinking practices. This taxonomy was created based on secondary data in the form of existing CT literature and CT instructional materials as well as primary data collected via interviews with mathematicians and scientists.

This research follows a similar approach to that documented in [3,8,11] where secondary data sources are analysed to develop new artefacts. The main aim of the research is to understand exactly what skills are associated with studying mathematics at a secondary school level and how these correspond to the skills addressed in CT. This understanding is documented in the form of a mapping of core CT skills to topics within the existing Mathematics curriculum for grades 10–12.

3 Definition of CT Skills Applied in This Research

The latest South African secondary school curriculum change happened in 2012 with the introduction of the National Curriculum and Assessment Policy Statement (CAPS). According to the Department of Basic Education, CAPS was not a new curriculum, but an amendment to the previous curriculum, the NCS, which introduced outcomes-based education in 2005 [9].

The NCS was based on the core principles of social transformation (ensuring that the educational injustices of the past are corrected and that there are equal educational opportunities for the entire population); active and critical learning (encouraging an active and critical approach to learning); and valuing indigenous knowledge systems (acknowledging the rich history and heritage of SA as important contributors to nurturing the values contained in the Constitution). According to the curriculum and assessment policy statement for grades 10–12 released in 2011, the new CAPS curriculum further aims to ensure that children are able to apply their acquired knowledge and skills in a way that is meaningful to their own lives [9].

Based on the work in [1] the nine CT concepts described below, were used to analyse the South African secondary school mathematics curriculum.

Data Collection (DC) and *Data Analysis* (DA): Data collection and data analysis are jointly defined in [8] as the ability to observe and describe phenomena and goal-oriented experiments. This includes the learner having the ability to collect measurement data and assess the data. In terms of mathematics, DC occurs when the learner is able to find a data source for a problem and DA occurs when the learner needs to count the occurrences of flips, dice throws and analyse these results.

Data Representation (DR): Data representation is defined as the ability to search, analyse and evaluate relevant information and data in different sources [8]. This ability includes being able to depict and organise data in appropriate graphs, charts or text. Examples of how this skill is used in mathematics are provided in [1], together with the explanation that DR is used when representing data in terms of histograms, pie charts and bar charts or when containing data in sets or lists.

Problem Decomposition (PD): Problem decomposition is described as a learner's ability to represent and divide a task into smaller manageable parts or break down a large problem into smaller sub-problems [8]. Barr and Stephenson [1] state that this skill is applied in mathematics when the learner must apply some order to operations in an expression.

Abstraction (AB): Abstraction is described as the ability to handle complex data and think in terms of conceptual ideas rather than merely in terms of the details [12]. This helps reduce the complexity of the problem as it assists the problem-solver to decide what to emphasize and what to hide. In mathematics, abstraction is used when variables are applied in algebra or essential facts are identified in a word problem and iterations are then used to solve the word problem [1].

Algorithms and Procedures (AP): Nyugen-thinh and Pinkwart [8] describe algorithms and procedures as the ability to develop a step-by-step solution to a problem or to follow a set of rules in order to solve a problem, while in [1], the authors explain that in mathematics AP is used when procedures such as long division, factoring or even addition and subtraction are performed.

Automation (AU): Automation is defined as the use of a computer or machine to do repetitive or tedious tasks [1]. In mathematics, AU is used when tools, such as Python programs, Sketch Pad or calculators, are applied to solve a problem.

Parallelization (PA): Parallelization is defined as the task of using a set of instructions to make several events occur simultaneously or organising resources to simultaneously carry out tasks to reach a common goal [1]. The authors explain that PA is applied in mathematics when solving linear systems or carrying out matrix multiplication.

Simulation (SI): Simulation is defined as the representation or model of a process and is applied in mathematics when functions are graphed on a Cartesian plane [1].

The next section illustrates how the skills described above are mapped to the mathematics curriculum for grades 10–12.

4 Mapping of CT Skills to the South African Mathematics Curriculum

Each objective within the ten topics of the NCS mathematics curriculum, namely functions, number patterns, finance, algebra, differential calculus, probability, euclidean geometry, trigonometry, analytical geometry and statistics, as provided by the Department of Basic Education, was analysed separately in this study. Table 1 summarizes the results obtained per topic. In addition, the analysis of one or more objectives in four of the topics, namely functions, algebra, differential calculus and Euclidean geometry, is presented in detail in the subsequent subsections.

Table 1. Summary of skills in the mathematics curriculum

	DC	DA	DR	PD	AB	AP	AU	PA	SI
Functions		X	X	X	X	X	X		X
Number patterns		X	X	X		X			
Finance		X	X	X	X	X			X
Algebra		X	X	X	X	X	X	X	X
Calculus		X	X	X	X	X	X		X
Probability		X	X	X		X			X
Euclidean geometry	X	X		X	X	X		X	
Trigonometry		X	X	X	X	X			X
Analytical geometry		X	X	X	X	X			
Statistics		X	X	X		X	X		X

Each analysis subsection begins with an explanation of the topic, followed by the objectives of the topic defined in a table along with the skills employed within that objective. Following this, one or two representative objectives are defined in detail, and a mathematical problem example (taken from one of the prescribed text books or past exam papers) is provided and analysed to show which CT skills are used during each of the steps carried out to solve the given problem.

4.1 Functions

A Grade 10 Mathematics textbook [5] defines functions as the mathematical building blocks for designing machines, predicting natural disasters, understanding world economies and keeping aeroplanes in the air. Functions take an input from many variables but always give the same output, which is unique to that function. Functions allow for the visualisation of relationships in terms of graphs that are easier to interpret than lists of numbers.

Table 2. Learning objectives and associated CT skills for the functions topic

	Learning objective	CT skills
1	Work with relationships between variables in terms of numerical, graphical, verbal and symbolic representations of functions and convert flexibly between these representations (tables, graphs, words and formulae)	DA, PD, DR, AB, SI
2	Generate as many graphs as necessary, initially by means of point-by-point plotting, supported by available technology, to make and test conjectures and hence generalise the effect of the parameter which results in a vertical shift and that which results in a vertical stretch and/or a reflection about the x-axis	DA, DR, PD, SI, AP, AU

The two learning objectives and corresponding skills used to achieve these objectives are given in Table 2. An analysis of the second objective is given below together with a worked example.

Objective 2 involves learners plotting graphs by firstly using a point-by-point plotting method and then using available technology (typically a Casio FX-82 ES Plus scientific calculator) to make and test conjectures. The other sub-objective is to test the effect of shifting the graphs both vertically and horizontally as well as reflecting them across the x- and y-axes.

The method of point-by-point graph plotting relies heavily on DR as learners are required to create a table of x and y variables that can then be used to plot a graph on a Cartesian plane, making use of SI. Decomposition occurs when the learner first finds the coordinates and then draws the Cartesian plane, identifying the different axes and then plotting the points. This process can also be considered as AP as plotting the graph is a process itself as the plane first needs to be set up and thereafter plotting can occur. The key skill used when striving to achieve this objective is simulation as this involves graphing a function in a Cartesian plane and modifying values of the variables. When the learner uses a calculator to create a table of variables they employ AU, which makes the point-by-point plotting more efficient. The act of shifting the graph and translating it employs AB.

Consider the function: $f(x) = \dfrac{-6}{x-3} - 1$

5.1.1   Calculate the coordinates of the x-intercept of f.

5.1.2   Calculate the coordinates of the y-intercept of f.

5.1.3   Sketch the graph of f in your ANSWER BOOK, showing clearly the asymptotes and the intercepts with the axes.

Fig. 1. Example 1 functions

The problem given in Fig. 1 requires the learner to interpret the equation and present it as a graph. In this example, the question has already been broken down into sub-questions for the learner instead of just asking for the third sub-question. If this were not the case, then the learner would have had to rely heavily on PD. Questions 5.1.1 and 5.1.2 in Example 1, rely on AP as the learner must recall that for an x intercept, the y variable is zero and for the y intercept, the x variable is zero. In order to solve Question 5.1.3 the learner can use the point-by-point method or a calculator to determine various points on the graph. Due to the complexity of the graph, however it is advised that the learner should have a rough idea of what the graph looks like, so some recall is also required. Decomposition occurs when the learner breaks up the problem by first finding the points and then plotting them, which employs SI.

Figure 2 illustrates a worked solution for Example 1. The first step is to analyse the question in order to understand how to proceed and which procedure to follow. This analysis is denoted by ① in the figure and shows where the DA skill is used. The first two sub-questions made use of the AP skill (denoted by ②) to find the x and y intercepts. The third sub-question requires the use of PD

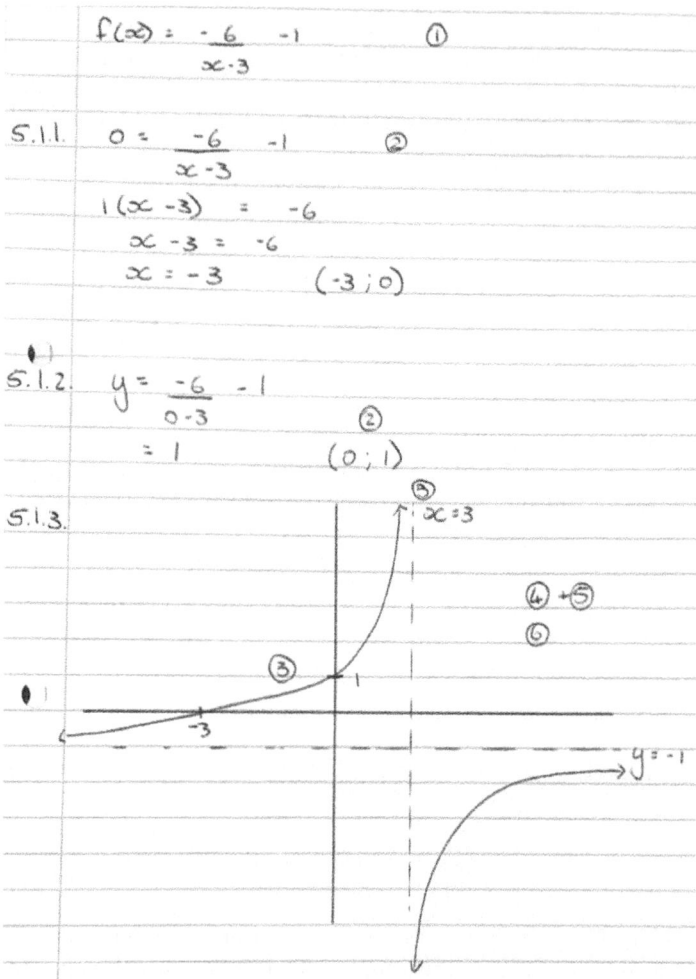

Fig. 2. Worked solution for Example 1

(denoted by ③) to first depict the intercepts and asymptotes on the Cartesian plane and then of SI (denoted by ④) to draw the graph. The data has been represented in the form of a graph and this employs the DR skill (denoted by ⑤). AU (denoted by ⑥) might be used if the points are calculated using the function mode on a calculator.

4.2 Algebra

Throughout history, individuals from many diverse cultures have contributed to the field of mathematics and although topics like algebra may seem obvious now, for many centuries mathematicians had to make do without this knowledge [5].

This section aims to explore more advanced and abstract mathematics that can be applied to everyday life.

Table 3. Learning objectives and CT skills for the Algebra topic

	Learning objective	CT skills
1	Describe the nature of roots, understanding that there are real and non-real numbers	PD, AP, AU, DA
2	Simplify expressions using the exponent laws for rational exponents	PD, AP, AB
3	Demonstrate an understanding of the definition of a logarithm and any laws needed to solve problems	AP, PD
4.	Manipulate algebraic expressions by: – Multiplying a binomial by a trinomial – Factorising polynomials up to the third-degree – Factorising the difference and sum of two cubes – Factorising by grouping in pairs	PD, AP, AB, AU
5.	Solve: – Linear equations – Quadratic equations – Quadratic inequalities in one variable and interpret the solution graphically – Exponential equations – System of linear equations – Word problems	PD, SI, AB, AP, AU, PA, DA, DR

Five learning objectives and the corresponding skills used to achieve these objectives are given in Table 3. An analysis of the last objective is given below along with two worked examples.

Objective 5 requires the learner to solve numerous equations including linear, quadratic and simultaneous equations as well as expressing an equation graphically and solving word problems, where the latter is only covered in grade 10. The simplest equations that learners will need to solve are linear equations and these can be solved using AP as the learner will follow a simple procedure. In order to solve quadratic equations, learners will use either a calculator, thereby applying AU, or the quadratic formula which implies using the AP skill. Quadratic inequalities are solved in numerous ways, but the learner will usually find the roots of the expression and may find it useful to sketch the graph in order to help describe the inequality, which makes use of PD and SI.

To find the roots of the quadratic equation the learner may use either a calculator or the quadratic equation, thereby using AU or PA, respectively. To solve simultaneous equations or systems of linear equations, learners make use of AP as there is a procedure to solve these equations. The learner begins by ensuring that the equations are in the same form and then proceeds to solve these either manually or using a calculator. Therefore, PD occurs, and AU may occur if the learner uses a calculator to solve the simultaneous equations. Due to the fact that the learner is solving two equations at the same time, PA occurs.

This objective also includes word problems for grade 10 learners. Solving these word problems relies on four key skills: DA, AB, AP and PA. The first two are used when the learner analyses the word problem and highlights the relevant and irrelevant information. PD is used to simplify the word problem by breaking it up into smaller parts that can then be solved using the AP skill.

Solve for x:

$$2x^2 - 2 \leq 3x$$

Fig. 3. Example 2 Algebra

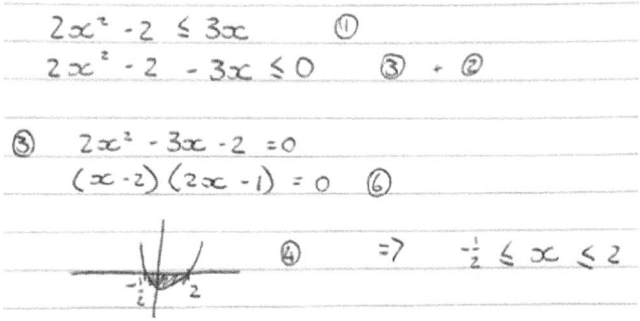

Fig. 4. Worked solution for Example 2

The problem depicted in Fig. 3 requires the learner to solve a quadratic equation with an understanding of inequalities. The learner needs to decompose the problem to do it in steps. The learner must first rearrange the equation and factorise it, which will require AP and PD skills. To solve the equation, the learner may use a calculator to find the solutions (using AU) or recall the laws of factorisation using the AP skill. Some learners may find it useful to plot a rough graph using the SI skill in order to describe the inequality. The learner will represent the data using an inequality.

Figure 4 illustrates the solution to Example 2. As an initial step, the expression is analysed using DA (denoted by ① in the figure) and the question decomposed by putting it in a form that can be factorised (denoted by ③). Now, the expression is factorised using AP (denoted by ②) or AU (denoted by ⑥). After finding the solutions, the problem is once again decomposed and an illustration is drawn to help visualize the inequality, which invokes SI (denoted by ④).

A second problem, depicted in Fig. 5, requires the learner to solve the two equations simultaneously in order to find x and y. The key skill used in this problem is parallelization as the learner has two equations to focus on at the same time. To solve this example, the learner needs to decompose the question and break it up into manageable parts. For example, the learner can look at each equation individually and try to simplify these. There are many ways that the learner can complete this problem. Using AB allows the learner to choose the most appropriate or easiest method and ignore the alternatives. The learner may opt to substitute the smaller equation into the quadratic equation to work out one of the variables. This will lead to a quadratic formula that can be factorised using a calculator (making use of AU) or by recalling the factorisation laws (thus making use of AP).

Solve simultaneously for x and y:

$$2 + y = -2x$$
$$-2x^2 + 8xy + 42 = y$$

Fig. 5. Example 3 Algebra

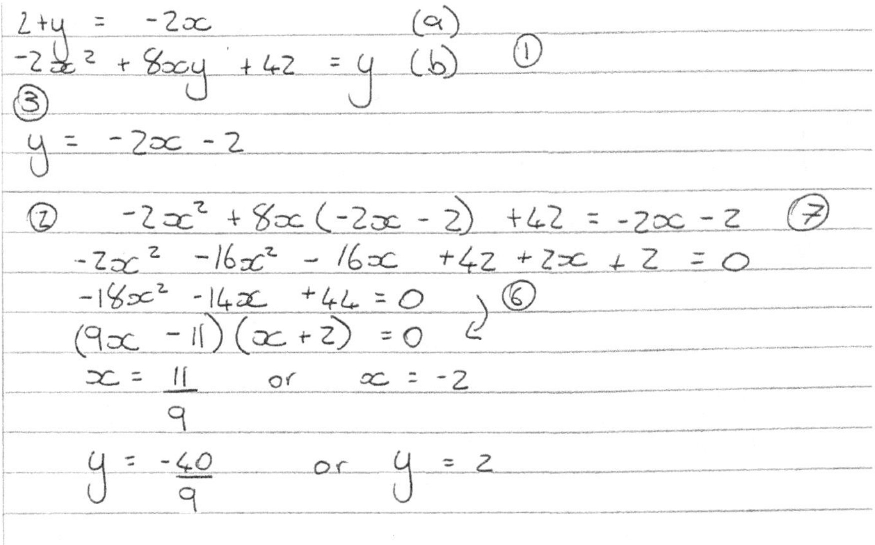

$$2 + y = -2x \quad (a)$$
$$-2x^2 + 8xy + 42 = y \quad (b) \quad ①$$
$$③$$
$$y = -2x - 2$$

② $\quad -2x^2 + 8x(-2x - 2) + 42 = -2x - 2 \quad ⑦$
$$-2x^2 - 16x^2 - 16x + 42 + 2x + 2 = 0$$
$$-18x^2 - 14x + 44 = 0 \quad ⑥$$
$$(9x - 11)(x + 2) = 0$$
$$x = \frac{11}{9} \quad \text{or} \quad x = -2$$

$$y = \frac{-40}{9} \quad \text{or} \quad y = 2$$

Fig. 6. Worked solution for Example 3

Figure 6 illustrates the worked solution to the simultaneous equations given as Example 3. The initial step is analysing the expression using DA (denoted by ①) in the figure). Following this the expression is decomposed (denoted by ③)) to rearrange the equation labelled a. This equation can then be substituted into the equation labelled b using PA (denoted by ⑦) and factorisation done by either recalling appropriate laws, thereby invoking AP (denoted by ②) or using a calculator, thereby invoking AU (denoted by ⑥).

4.3 Differential Calculus

Table 4. Differential calculus learning objectives and associated CT skills

	Learning objective	CT skills
1	Differentiation of specified functions from first principles	PD, AP
2	Use of the specified rules of differentiation	AP, PD
3	The equations of tangents to graphs.	PD, AP, AB, DA
4	The ability to sketch graphs of cubic functions	DA, SI, PD, AP, DR, AU
5	Practical problems involving optimization and rates of change (including the calculus of motion)	PD, DA, AB, AP

Calculus is a central branch of mathematics that developed from algebra and geometry and consists of two related ideas, namely, differential calculus and integral calculus. The curriculum only focuses on differential calculus and explores how it can be used in everyday optimisation problems and finding rates of change, amongst others [6].

A cubic function f has the following properties:
- $f\left(\dfrac{1}{2}\right) = f(3) = f(-1) = 0$
- $f'(2) = f'\left(-\dfrac{1}{3}\right) = 0$
- f decreases for $x \in \left[-\dfrac{1}{3}; 2\right]$ only

Draw a possible sketch graph of f, clearly indicating the x-coordinates of the turning points and ALL the x-intercepts.

Fig. 7. Example 4 differential calculus

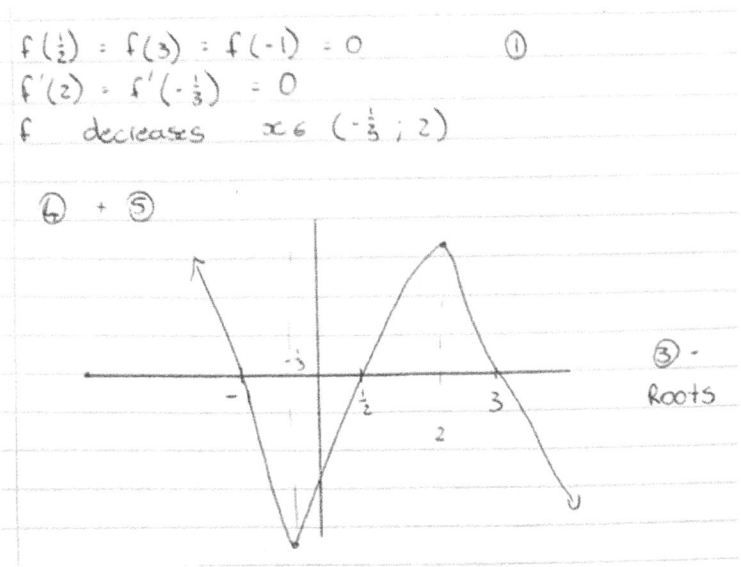

Fig. 8. Worked solution for Example 4

Five learning objectives and the corresponding skills used to achieve these objectives are given in Table 4. An analysis of the fourth objective is given below along with a worked example.

Within learning objective 4, learners are required to sketch cubic graphs using differential calculus. There are a number of different ways to approach this, but the key to solving these problems is PD as learners need to be able to break up a problem into manageable parts. The procedure that the learner will follow is based on the information provided, and therefore, AP will be applied differently depending on what information is available. To find the roots of a cubic function learners may use a calculator or their factorising knowledge, and thus, this step may involve AU or AP, respectively. DR and SI are invoked to sketch the graph, after the roots, turning points and intercepts have been calculated.

The problem depicted in Fig. 7 requires the learner to sketch a cubic graph based on the information provided. Since the learner has not been provided with the equation of the cubic function, the graph cannot be plotted using the function mode on the calculator. Before attempting the question, the learner must analyse the data that has been given (using DA) to evaluate whether there is adequate information to solve the problem and this will also allow the learner to decide which procedure to follow. The learner must decompose the problem using PD, which helps to deal with the complexity of the problem. The learner must understand what each of the points means, for example, the first derivative equal to zero denotes a turning point and thus, the turning points are given. This means that the learner has five points, which is sufficient to determine the concavity of the functions. At this stage, the learner can use DR and SI to plot the cubic function.

Figure 8 illustrates a worked solution to Example 4. The initial step involves analysing the problem using DA (denoted by ① in the figure) to establish what data has been given to sketch the cubic function. This data must be represented in the form of a graph, which uses DR (denoted by ⑤). The sketching process can be decomposed using PD (denoted by ③); this is done by first establishing where the roots lie on the Cartesian plane as well as the turning points. The last line of information helps describe the concavity of the graph. Following this the graph can be sketched using SI (denoted by ④).

4.4 Euclidean Geometry

According to a Grade 10 Mathematics textbook [5], geometry developed as the area of knowledge that deals with spatial relations. Euclidean Geometry deals with space and shapes using a system of logical deductions and can be applied in real life to surveying and designing new buildings, amongst others.

Two learning objectives and the corresponding skills used to achieve these objectives are given in Table 5. An analysis of the first objective is given below along with a worked example.

Objective 1 involves analysing a circle to deduce the problem and find angles using the rules of circle geometry. The first step in solving these problems is to analyse the data given to establish what angles or chords still need to be found. The learner will need to decompose the question as it may be necessary to find more than one angle or chord. Finding these angles invokes the DC skill. The learner may have to use abstraction to ignore angles that are not labelled and do not need to be found. Parallelization occurs when the learner must find multiple angles at once.

Table 5. Euclidean Geometry learning objectives and associated CT skills

	Learning objective	CT skills
1	Investigate and prove theorems of the geometry of circles as well as solving circle geometry problems, providing reasons for statements when required	AP, AB, PD, PA, DC
2.	Prove: – A line drawn parallel to one side of a triangle divides the other two sides proportionally (and the Mid-point Theorem as a special case of this theorem) – Equiangular triangles are similar – Triangles with sides in proportion are similar – Pythagorean Theorem by similar triangles	AP, PD, DA

Determine the values of h and s.

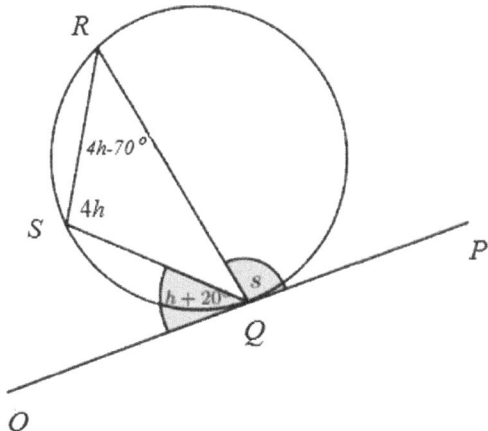

Fig. 9. Example 5 Euclidean Geometry

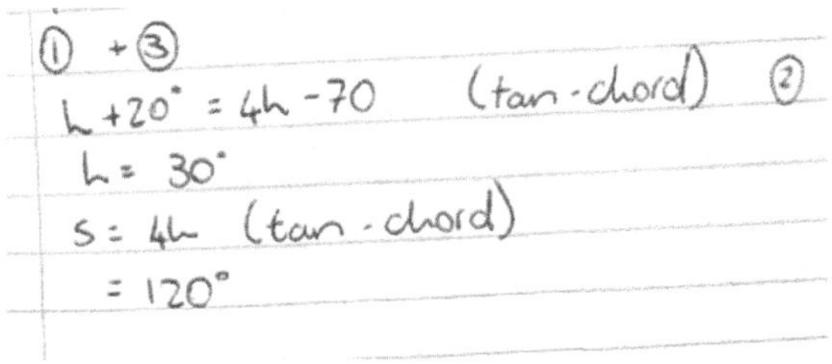

① + ③
$h + 20° = 4h - 70$ (tan-chord) ②
$h = 30°$
$s = 4h$ (tan-chord)
$\quad = 120°$

Fig. 10. Worked solution for Example 5

The problem depicted in Fig. 9 requires the learner to analyse the circle and the information on the circle and use this information to recall the rules of circle geometry to find the variables h and s. The first step in solving the problem is to use DA to analyse the information given. The learner may opt to use PD to break up the question and find the angles individually or may use PA to find them concurrently. Finding the angles invokes the DC skill while the learner uses AB to ignore any angles that do not need to be found. The next step involves recalling and applying the tan-chord rule as well as the use of algebra to find the variables, which requires the AP skill.

Figure 10 illustrates the worked solution to Example 5. The initial step involves analysing the problem using DA (denoted by ① in the figure) to establish what data has been provided in any illustrations or text in the question. Following this the problem is decomposed (denoted by ③) to first find h, using AP (denoted by ②) to recall the rules of circle geometry and apply the tan-chord theorem. After h has been found, AP is again used to apply the tan-chord theorem to find s. DC is used when the learner finds the angles as this can be seen as a form of collecting data.

5 Conclusion and Future Work

In this paper, we provided some insight into the full exploration undertaken whereby nine key CT skills identified from the literature, were mapped to the ten topics in the South African Grades 10–12 Mathematics curriculum. Examples of problems from some of the topics in the curriculum were presented together with step-by-step worked solutions showing which of the CT skills are invoked in solving these problems.

The final mapping of skills to representative problems for all objectives within all topics shows that none of the skills is unused in the mathematics curriculum, although DC, AU and PA are minimally used. However, development of these skills can be enhanced with the inclusion of various additional exercises in the curriculum. To improve DC, practical experiments could be added to the curriculum so that learners have the opportunity to physically collect data, while the use of AU can be enhanced by incorporating computers in mathematics using software such as Microsoft Excel to automate procedures such as plotting functions. PA is the hardest skill to incorporate further in the curriculum as it requires the introduction of matrices, which are considered too complex for a secondary school curriculum.

Having identified the skills associated with each topic in the curriculum, future research will focus on the reciprocal relationship between CT and mathematics with the design of CT exercises or interventions, not necessarily based on actual mathematical content. For example, activity worksheets aimed at a specific problem solving need, or physical puzzle building or other activities could be used to develop and/or enhance the core CT skills, which in turn could assist learners that struggle with comprehending the core mathematics curriculum. It is envisaged that such interventions would not only develop better mathematical skills and general problem-solving skills amongst the school learners, but also equip them with the requisite basic skills to succeed in undergraduate CS courses.

References

1. Barr, V., Stephenson, C.: Bringing computational thinking to K-12: what is involved and what is the role of the computer science education community? ACM Inroads **2**, 49–54 (2011). https://doi.org/10.1145/1929887.1929905
2. Beaubouef, T., Mason, J.: Why the high attrition rate for computer science students: some thoughts and observations. SIGCSE Bull. **37**(2), 103–106 (2005)
3. Howland, K., Good, J., Nicholson, K.: Language-based support for computational thinking. In: 2009 IEEE Symposium on Visual Languages and Human-Centric Computing (VL/HCC), pp. 147–150. IEEE, Corvallis (2009). https://doi.org/10.1109/VLHCC.2009.5295278

4. Jackson, J., Moore, L.: The role of computational thinking in the 21st century. In: Proceedings of 5th International Conference on Appropriate Technology, pp. 149–155 (2010)
5. Kannemeyer, L., Jenkin, A., van Zyl, M., Scheffler, C.: Everything Maths Grade 10 Mathematics, vol. 1. Siyavula Education (2012)
6. Kannemeyer, L., Jenkin, A., van Zyl, M., Scheffer, C.: Everything Maths Grade 12 Mathematics, vol. 1. Siyavula Education (2012)
7. Mwakapenda, W.: Understanding connections in the school mathematics curriculum. South Afr. J. Educ. **28**, 189–202 (2008)
8. Nguyen-thinh, L., Pinkwart, N.: K-12 computational thinking education in Germany. In: International Conference on Computational Thinking Education 2017, pp. 39–43. The Education University of Hong Kong (2017)
9. du Plessis, E., Marais, P.: Reflections on the NCS to NCS (CAPS): foundation phase teachers' experiences. Indep. J. Teach. Learn. **10**, 114–126 (2015)
10. Voskoglou, M., Buckley, S.: Problem solving and computers in a learning environment. Egyp. Comput. Sci. J. **36**, 28–46 (2012)
11. Weintrop, D., et al.: Defining computational thinking for mathematics and science classrooms. J. Sci. Educ. Technol. **25**(1), 127–147 (2015). https://doi.org/10.1007/s10956-015-9581-5
12. Wing, J.M.: Computational thinking. Commun. ACM **49**(3), 33–35 (2006). https://doi.org/10.1145/1118178.1118215
13. Wing, J.M.: Research notebook: computational thinking - what and why. Link Mag. pp. 20–23 (2011)
14. Zhang, N., Biswas, G.: Defining and assessing students' computational thinking in a learning by modeling environment. In: Kong, S.-C., Abelson, H. (eds.) Computational Thinking Education, pp. 203–221. Springer, Singapore (2019). https://doi.org/10.1007/978-981-13-6528-7_12

Common Code Writing Errors Made by Novice Programmers: Implications for the Teaching of Introductory Programming

Mokotsolane Ben Mase[1] and Liezel Nel[2]

[1] Department of Computer Science and Informatics, University of the Free State, Phuthaditjhaba, South Africa
masemb@ufs.ac.za
[2] Department of Computer Science and Informatics, University of the Free State, Bloemfontein, South Africa
nell@ufs.ac.za

Abstract. Novices tend to make unnecessary errors when they write programming code. Many of these errors can be attributed to the novices' fragile knowledge of basic programming concepts. Programming instructors also find it challenging to develop teaching and learning strategies that are aimed at addressing the specific programming challenges experienced by their students. This paper reports on a study aimed at (1) identifying the common programming errors made by a select group of novice programmers, and (2) analyzing how these common errors changed at different stages during an academic semester. This exploratory study employed a mixed-methods approach based on the Framework of Integrated Methodologies (FraIM). Manual, structured content analysis of 684 programming artefacts, created by 38 participants and collected over an entire semester, lead to the identification of 21 common programming errors. The identified errors were classified into four categories: syntax, semantic, logic, and type errors. The results indicate that semantic and type errors occurred most frequently. Although common error categories are likely to remain the same from one assignment to the next, the introduction of more complex programming concepts towards the end of the semester could lead to an unexpected change in the most common error category. Knowledge of these common errors and error categories could assist programming instructors in adjusting their teaching and learning approaches for novice programmers.

Keywords: Novice programmer · Common programming errors · CS1 · Computer Science education

1 Introduction

Novice programmers face various challenges when learning to write computer programs [38]. These programming challenges have been shown to contribute to high failure and dropout rates in computer programming courses [8]. Novices often end up being unable to write code properly due to a lack of problem-solving strategies and fragile knowledge of basic programming concepts [13]. Numerous studies have been conducted to identify the difficulties experienced by novice programmers and the common programming errors made by them [7, 33]. According to [35], it is key that instructors know which programming errors their students are making. Better insight regarding the specific errors made can help instructors to formulate custom teaching and learning strategies specifically aimed at addressing the programming challenges of their students. This paper, therefore, attempts to answer the following two questions:

- What are the most common code writing errors made by a select group of novice programmers?
- How do these errors change during an academic term?

The remainder of this paper is structured as follows. Section 2 presents relevant background literature. This is followed by a discussion of the research method in Sect. 3 and a presentation of the results and discussion in Sect. 4. The paper concludes with conclusions and recommendations in Sect. 5.

2 Background

Programming is a fundamental part of the Computer Science (CS) curriculum. However, novice programmers often perceive programming to be complex [38]. Teaching computer programming skills within the university context has also proven to be a difficult task for instructors [26]. Over the years, instructors have shown growing concern over the difficulties experienced by novices when learning programming principles [33]. Since the goal of an introductory programming course is to train novices, it is expected that at the end of their first semester, they should be able to write simple programs [33]. However, [39] notes that even after two years of learning to program, most novices still struggle to write code properly. Since novices are inexperienced in the art of programming, they often lack the necessary programming knowledge and strategies to solve problems [29]. [38] point out that novices encounter specific difficulties when they write code due to the abstract nature of the programming language and a lack of problem-solving skills in their early stages of learning to program. Due to their lack of problem-solving skills, novices are unable to learn the theoretical concepts of the programming domain and are therefore unable to practice these concepts in designing different programs [26]. Consequently, novices end up committing various common programming errors when they write computer code.

2.1 Common Programming Error Categories

While learning to program, novices are bound to make unnecessary errors. Correcting these errors can be a time-consuming process especially if the novice does not understand the compiler error messages [28]. [7] concur that novices often find it challenging and frustrating to understand feedback presented through compiler error notifications as some notifications are imprecise and confusing. To further complicate error correction, not all programming errors are picked up by the compiler. Numerous studies have been conducted to gain an in-depth understanding of the kinds of programming errors novices are likely to make [22, 32]. Programming errors are typically classified into four categories: syntax, semantic, logical, and type errors.

Syntax Errors. A syntax error occurs when the grammatical rules of the programming language being used are not followed [22]. This type of error can be a major obstacle for novices as it is likely to slow down their progress [14]. The correction of syntax errors is usually the first step in the debugging process since it is impossible to continue with program development until these errors have been dealt with [14]. One of the major reasons why novices fail to master the art of programming relates to their inability to apply valid rules of syntax when writing a program [29].

Semantic Errors. This type of error occurs when the meaning of the programming code is not consistent with the programming language [22]. Examples include the use of a non-initialised variable or an attempt to divide a value by zero. Such errors are violations of the semantic meaning because it deals with the interpretation of the programming language [10]. Semantic errors typically occur due to the misuse of a programming concept, despite correct syntactical structure, and are detected when the program code is compiled [39].

Logic Errors. Logic errors occur when the program does not solve the given problem [16]. For instance, if the written code instructs the compiler to execute command A under condition X while the command should instead be executed under condition Y to solve the given programming problem, this will result in a logical error. Since this type of error is typically detected when the program has successfully compiled but fails at the time of execution, it can be harder to find and debug than the other error types [34]. Consequently, logic errors are generally only detected and addressed through rigorous software testing.

Type Errors. Type error occurs when the indicated types are incorrect [1]. For example, attempts to assign a floating-point value to an integer location. [31] add that the type error can be detected when a novice programmer tries to use variable values with different data type or perform operations that are not allowed by the variable type.

Summary. According to [25], novices spend more time trying to understand programming language syntax and semantics while paying less attention to problem-solving strategies. If novices are made aware of the different kinds of programming errors they typically make, that could ultimately help to improve their understanding of programming [32]. Knowledge regarding typical programming errors made by their students can also assist programming instructors in making informed decisions regarding the design of their teaching and learning tools and strategies [3].

2.2 Strategies Utilized in the Identification of Common Programming Errors

The monitoring of student programming behavior and mistakes has a long history in CS education research [4]. Numerous studies have attempted to monitor and identify the most common errors made by novice programmers [4, 16]. In these prior studies, a variety of strategies have been followed to identify common programming errors.

Using self-reported survey data collected from undergraduate students, teaching assistants and CS educators, [19] compiled a list of 62 typical errors made by novice programmers in Java courses. A process of data refinement and elimination (based on the overall aim of their study) was then used to reduce the original list of errors to 20. Sixty-five percent of the errors from their final list are syntax errors.

In the study conducted by [28], data was firstly collected through the BlueJ integrated development environment (IDE) for programs developed in the Java programming language. The Blackbox project data was also used and 23 sessions comprising a total of 136 events recorded were randomly selected for analysis. They identified a total of 80 error categories that were divided into syntax, semantic, and logic errors. According to their findings, semantic errors are more likely to correspond with compiler error messages than syntax errors.

In their study, [4] investigated novice programming mistakes by using a year's worth of compilation events from more than 250,000 novices spread all over the world. The data for their study was sourced from the large Blackbox data set that collects Java code of BlueJ users. They used 18 mistakes, identified in their earlier work [12], as the basis of their analysis. A comprehensive empirical analysis of the time-to-fix and frequency of different Java errors was provided. [4] present their findings in three error categories: syntax, semantic and type errors. They identify semantic errors and type errors as the most common.

[6] investigated common errors leading to excessive student struggles with homework problems. They analyzed 78 auto-graded coding homework problems developed in the C++ programming language. They found the struggle rate to be 10–15% and as high as 30–40% for some coding activities. The researchers then investigated the errors that led to these struggles by manually examining the student submissions for the 10 homework problems with the highest struggle rates. Their results show that the most common errors occur in nested loops, `else-if` statements, random range, input/output, `for` loop and vector, `for` loop and `if`, vector index, negated loop expressions, and Boolean expressions.

[31] carried out a study where they implemented a data-driven approach to identify common errors in Java using the data from an automated assessment tool called Mulberry. Among the 766 compilation-erroneous solutions, 1146 compilation errors existed. By grouping the same compilation errors, 28 distinct compilation errors were identified and 15 of them were categorized as common errors. In the same year, [2] attempted to identify common errors from C programs written by novices. The 12371 submissions were inspected manually. The identified errors were classified into six categories: syntactic, conceptual, strategic, sloppiness, misinterpretation, and domain knowledge. Interestingly, their findings show that a big portion of errors made by novices is simply caused by sloppiness.

In most of the prior studies mentioned here, the common errors were identified using different software tools or compiler error messages that could easily be evaluated automatically. However, [28] warn that analysis of compiler error messages is imprecise. Firstly, a single error can produce different compiler error messages. Secondly, the same compiler error message can be produced by entirely different errors, depending on the context. Lastly, the compiler error messages do not always have a direct correlation to the types of errors made by novices. Although [6] performed some manual analysis of programming code, the purpose of their manual process was to discover possible reasons for why the novices' might have made each of the already identified errors. The current study followed a different approach by performing structured content analysis as part of an extensive manual analysis of programming artefacts–as outlined in Sect. 3.

3 Method

In this exploratory study, a case study approach based on the Framework of Integrated Methodologies (FraIM) [30] was followed. The aim of this case study was firstly to identify and classify all the code writing mistakes made by a group of first-year CS students at a selected university. The sample comprised 38 novices registered for a first-year introductory programming course. The sampling strategy can be regarded as both purposeful [23] and convenient [17] since the selected novices were in the process of learning how to program and the principal researcher (the first author), as an instructor for this course, had direct contact with the novices.

For this case study, artefact analysis [30] was selected as the primary data collection strategy. The artefacts included all the pieces of C# programming code written by each of the students over an entire 13-week semester for 11 timed assessments–comprising eight practical assignments, two semester tests (ST1 and ST2) and the final examination. These artefacts comprised the actual programming files submitted by the students which were all created within the Microsoft Visual Studio IDE using the C# language. Microsoft Visual Studio 2019 code metrics were used to check the lines of code as well as the lines of executable code. Table 1 provides a summary of the programming artefact collected and analyzed for each assessment.

Each assessment included at least one but no more than four programming questions. Since all the students completed all the assessments, there were 38 artefacts for each of the 18 programming questions–resulting in a total of 684 artefacts. A structured content analysis approach was followed to analyze each of the 684 artefacts. First, a manual line-by-line review of each piece of program code was conducted to identify all the code writing mistakes made by each student. Where relevant, error messages and visual feedback provided by the IDE were also reviewed to assist in the identification of errors. A closed coding schedule was then applied to map each identified error to one of the 18 mistakes previously identified by [4]. However, it became apparent that some of the identified errors could not be classified according to the initial coding schedule. The coding schedule was therefore adapted by adding additional error descriptions for these unclassified errors. All the data was captured in Microsoft Excel and descriptive statistics were used to analyze the resulting numerical data.

Table 1. Programming artefacts

Assessment type	Tasks	No. of questions	Lines of code	Lines of executable code
Formative	Prac1	1	2293	931
	Prac2	1	2875	1144
	Prac3	1	2296	794
	Prac4	1	2945	1117
	Prac5	1	2062	590
	Prac6	1	2718	694
	Prac7	1	5110	1026
	Prac8	1	4022	637
Summative	ST1	3	7146	2436
	ST2	3	12646	2369
	Exam	4	12790	3106
Total		18	56903	14844

4 Results and Discussion

The discussions in the following sub-sections focus on the common errors identified from the students' coding artefacts as well as how these common errors changed at different stages of the academic term.

4.1 Common Programming Errors Made by Novices

Analyses of the artefact data revealed a total of 2396 errors made by the novices during the course of the semester. Of these 311 (12.98%) were syntax errors, 787 (32.85%) were type errors, 1108 (46.24%) were semantic errors and 190 (7.93%) were logic errors. A summary of the 21 identified errors is presented in Table 2 where errors are grouped by category. For each category, the errors are sorted according to the number of recorded occurrences (from high to low). To aid in consequent referencing to specific errors, each error is labelled with an alphabetical character. The identified errors corresponded with 12 of the 18 original errors from [4] while nine additional errors (see errors **D, F, G, H, P, R, S, T** and **U** in Table 2) were identified. The top ten errors (**A, C, B, D, E, F, L, M, R** and **N**) accounted for 87.65% (2100) of the total errors. Error **A** (with 565 occurrences; 23.58%) and error **C** (with 540 occurrences; 22.54%) were the two most common errors made by the novices. In the following sub-sections, we look closely at some of the most common errors from each of the categories listed in Table 2 and compare our results to those of previous studies.

Type Errors. While the type error "Calling method with wrong types" was among the top-18 identified errors [accounting for 393 (13.92%) of the total 2824 errors] in the study by [1], the similar type error "Invoking methods with wrong arguments" only accounted for 38 (1.11%) of the 3408 errors identified by [2]. In contrast, type error **A** (Invoking methods with wrong arguments) was the most frequent error [565 (23.58%) of the total 2396 errors] in our study. [15] point out that when a method body is implemented,

Table 2. Summary of student errors

Error Category	Error Description	No. of errors	% of overall errors	Overall rank
Type	A. Invoking methods with wrong arguments (e.g. wrong types).	565	23.58	1
	B. Incompatible types between method return and type of variable that the value is assigned to.	222	9.27	3
Semantic	C. Class claims to implement an interface but does not implement all the required methods.	540	22.54	2
	D. Invoking class method on object.*	159	6.64	4
	E. Control flow can reach end of non-void method without returning.	150	6.24	5
	F. Undefined constructor in implicit constructor call.*	135	5.63	6
	G. Use of a non-initialized variable.*	50	2.09	12
	H. Object not defined.*	36	1.50	15
	I. A method that has a non-void return type is called and its return value ignored/discarded.	20	0.83	16
	J. Trying to invoke a non-static method as if it was static.	13	0.54	18
	K. Use of == instead of .equals to compare strings.	5	0.21	19
Syntax	L. Including the types of parameters when invoking a method.	97	4.05	7
	M. Confusing short-circuit evaluators (&& and ||) with conventional logical operators (& and |).	91	3.80	8
	N. Confusing the assignment operator (=) with the comparison operator (==).	61	2.55	10
	O. Getting greater than or equal/less than or equal wrong (i.e. using => or =< instead of >= and <=.	54	2.25	11
	P. Cannot implicitly convert type int [] to string [].*	4	0.17	20
	Q. Wrong separators in for loops (e.g. using commas instead of semi-colons).	4	0.17	21
Logic	R. Method with parameters: Confusion between declaring parameters of a method and passing parameters in a method invocation.*	80	3.34	9
	S. Improper casting.*	47	1.96	13
	T. Array index out of range.*	46	1.92	14
	U. Invoking a non-void method in a statement that requires a return value.*	17	0.71	17
Total errors		2396		

*Additional error categories identified in this study.

novices should ensure that the type of the return expression matches the return type of the method. It is therefore extremely important that the novices understand that the types of arguments in a method call should match the types of the arguments in the method's definition [19]. In the study by [31], the error "Incompatible types", which is similar to our type error **B** (Incompatible types between method return and type of variable that the value is assigned to), accounted for 3.23% of the 1146 recorded errors. The same error only accounted for 0.79% of the errors identified by [2]. Even though the previous work shows low frequencies for error **B**, this error seems to be problematic in our case because it occurred frequently and accounted for 9.27% of the total errors. This type error appears when novices use a non-void method that returns a value that has a different data type to the variable that is supposed to receive the returned value [5]. Instructors must warn novices about this type of error so that they can avoid equating two variables of different types.

Semantic Errors. While error **C** (Class claims to implement an interface, but does not implement all the required methods) was identified as the second highest occurring error in our study [accounting for 540 (22.54%) of the recorded errors], errors with similar descriptions had much lower occurrences in the studies by [27] (0.20%) and [1] (0.04%). According to [21], error **C** occurs when novices forget to enclose their methods with a class. [39] adds that the class or interface expected error can also take place when there is code outside of a class declaration. This error usually occurs when objects declared at the start of the class are mistakenly placed before the actual beginning of the class. Similarly, we recorded a higher occurrence (6.64% of total errors) for error **D** (Invoking class method on object) than what [27] (1.99%) observed for a similar error in their study. Error **D** typically occurs when novices are trying to invoke a method that belongs to a class on a variable or an object directly [39]. The apparent less common nature of error **J** (Trying to invoke a non-static method as if it was static) (0.54% of total errors) in our study is in line with the findings of [1] and [27] who recorded occurrences of 1.98% and 1.59% respectively for a similar error. Error **J** takes place when novices use a non-static method as if it was static [36]. Error **K** (Use of == instead of .equals to compare strings) was also found to occur seldomly with novices in our study with only five occurrences (0.21%). Although more than half of the semantic errors identified in our study occurred relatively infrequently, four of the top-6 recorded errors were from the semantic error category. Novices tend to find it quite frustrating or difficult to resolve semantic errors, and this is mostly due to their lack of programming knowledge [20]. If instructors are aware of the semantic errors that are most problematic for novices, they could use this knowledge to devise teaching strategies aimed at reducing novices' programming frustrations.

Syntax Errors. Even though syntax errors are based on misspelling, punctuation and word order in a program, novices typically find it difficult to solve some of these errors. For instance, in the study by [27], the error "Method call: Parameter number mismatch", which is comparable to our error **L** (Including the types of parameters when invoking a method), accounted for 2.68% of the total errors. This moderately corresponds with the findings of our study with error **L** accounting for 4.05% of the total error occurrences. This error occurs when novices add types to parameters when they call a method that does not require types [5]. The results of [31] indicate an occurrence of 5.50% for the error "Incorrect use of operators", which is in a similar range as the 3.08% we recorded for our error **M** [Confusing short-circuit evaluators (&& and ||) with conventional logical operators (& and |)]. [1] point out that the error "Confusing = with ==" is problematic for novices [accounting for 262 occurrences (9.28%) from the total 2824 errors identified in their study]. However, for the same error (**N**) we recorded a much lower occurrence frequency of 2.55%, while [2] observed an even lower frequency of 0.50%. As programming instructors, we have often in the past came across instances where novice programmers use the assignment operator (=) instead of the equality operator (==) to compare values in an expression. It is possible that our prior awareness of this common problem could have led us to pay specific attention to this issue during our teachings. By raising awareness of this common syntax error, our students possibly have been better prepared to resolve this issue on their own.

Logic Errors. Since our initial coding schedule (adapted from [4]) did not include any logic errors, all the error descriptions in this category were adapted from other literature sources. While error **R** (Method with parameters: Confusion between declaring parameters of a method and passing parameters in a method invocation) was one of the top-10 errors identified in our study (with a frequency of 3.34%), [27] found their similar "Method call: Parameter type Mismatch" error to occur less frequently (0.99%). Novices experience this error when there is confusion between passing parameters, declaring them, and identifying them in the method's definition [11]. Another logic error that novices seem to find challenging is error **S** (Improper casting), which accounted for 1.96% of the recorded errors. Again, [27] recorded an even lower frequency of 0.10% for the same error. This error is experienced when novices declare variables rather than casting them [19]. According to [16], novices often display a lack of knowledge regarding casting integers to doubles. Since novices struggle to find logic errors, developing a better understanding of logic errors may help to inform teaching practices and reduce novices' frustration [16].

Summary. Due to a lack of programming knowledge, novices appear to find type errors and semantic errors the most difficult to detect and fix. [16] agree that novices can find it very frustrating to locate and resolve these errors. Since novices repeat these common errors, it could serve as a further indication that they continue to struggle with these errors [9]. Instructors should therefore pay special attention to how often these errors are repeated as the number of repeated occurrences could be a good indicator of how well or poorly novices are performing when they write code [37]. Instructors should also look more closely at each specific error because the more they understand the nature of these errors, the more effective their teaching strategies can be [15]. Novice programmers should be encouraged to find and correct errors by first understanding what the code actually does as compared to what it is supposed to do.

4.2 Changes in Common Errors During an Academic Term

As part of an efficient teaching and learning strategy for the teaching of programming, it is crucial that instructors monitor novices' progress throughout the entire academic term.

Monitoring of novices' learning processes could help to identify individuals who are at risk of failing the programming course [18]. Instructors often have the best understanding of the type of support their students need and how best to address the specific needs of novices [24]. The discussions in the following sub-sections look closely at the changes in common programming errors made by novices in the various formative and summative assessment activities throughout an academic term.

Formative Assessments: Most Common Errors. In the semester during which this study was conducted, the novices had to complete eight practical programming assignments. The graph in Fig. 1 depicts the number of occurrences for the most frequent error (from Table 2) as identified for each of these assignments.

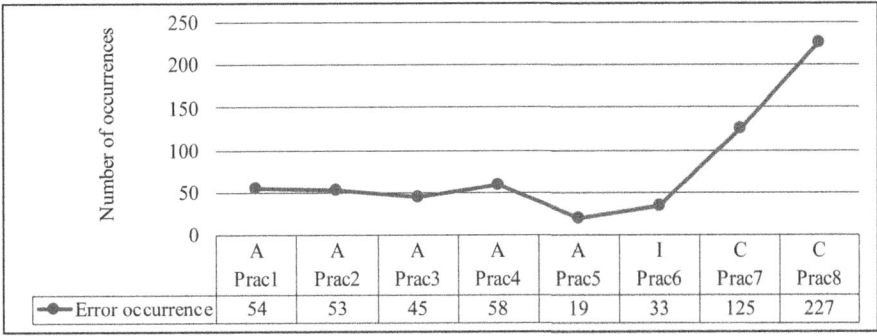

Fig. 1. Most common error for each formative assessment

Error **A** (Invoking methods with wrong arguments) was the most frequent error made by novices in the first five assignments (Prac1 to Prac5) while Error **I** (A method that has a non-void return type is called and its return value ignored/discarded) was the most common error during Prac6. However, error **C** (Class claims to implement an interface but does not implement all the required methods) was the most common error for both Prac7 and Prac8. It is notable how the occurrence of this error increased dramatically from Prac7 to Prac8 with 102 more occurrences in the final assignment. However, it should be noted that in Prac7, advanced concepts such as inheritance, polymorphism, abstract classes as well as abstract and virtual methods were introduced. The novices struggled with the implementation of these concepts because these assignments were the first that required knowledge of these advanced concepts. The prominence of error C in the final two assignments is further evidence of the novices' struggles to implement new (or unfamiliar) concepts. Overall, the results depicted in Fig. 1 show that the most common errors made by the novices during the formative assessments can be attributed to their lack of knowledge regarding the implementation of classes, interfaces and methods.

Formative Assessments: Occurrence of Error Categories. As outlined in Table 2, the identified errors were grouped into four categories. Figure 2 depicts the number of occurrences for these error categories in each of eight practical assignments. It is interesting to note how the most common error category change between the assignments.

While type errors were the most common category in four of the first five assignments, there was an unexpected surge in the occurrence of semantic errors towards the end of the semester (with 187 occurrences in Prac7 and 338 occurrences in Prac8). Although there are indications that the most common error category is likely to be the same from one assignment to the next, there is also a chance of another error category becoming more prominent due to the nature and complexity of the assignment. Since semantic errors deal with the meaning of the code and arise from mistaken ideas on how a programming language interprets instructions, instructors should pay more attention to how novices can be supported in this regard [40], especially when more complex concepts are introduced in later parts of the academic term.

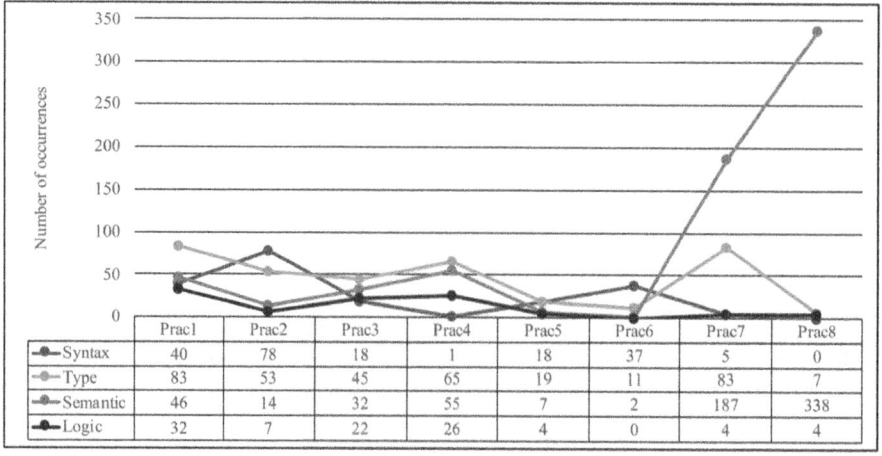

Fig. 2. Error categories for formative assessments

Summative Assessments: Most Common Errors. The novices had to complete two practical semester tests (ST1 and ST2) and a practical examination (Exam) as part of their summative assessments. The graph in Fig. 3 depicts the number of occurrences for the most frequent error (from Table 2) identified for each of these assessments. Error **A** (Invoking methods with wrong arguments) was the most frequent error made by novices in ST1 while Error **C** (Class claims to implement an interface but does not implement all the required methods) was the most common error in ST2. However, error **B** (Incompatible types between method return and type of variable that the value is assigned to) was the most frequent error during the Exam (with 132 occurrences). Overall, the most common errors made by the novices in the summative assessments can be attributed to their lack of knowledge regarding the implementation of return types and values.

Summative Assessments: Occurrence of Error Categories. In the case explored in this study, ST1 took place after the completion of Prac5 while ST2 took place after Prac8. ST2 was followed by the Exam at the end of the academic term. Figure 4 depicts

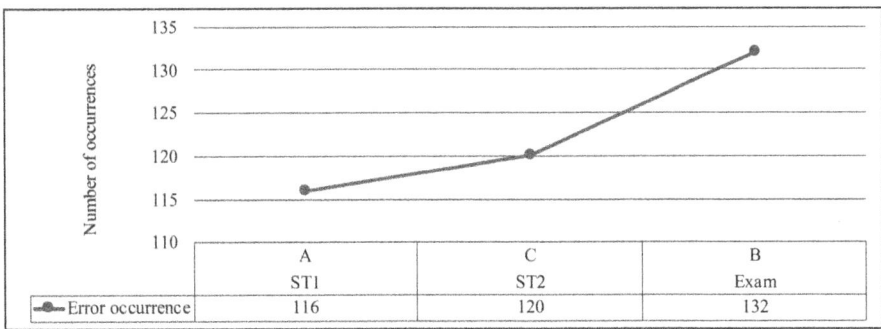

Fig. 3. Most common error for each summative assessment

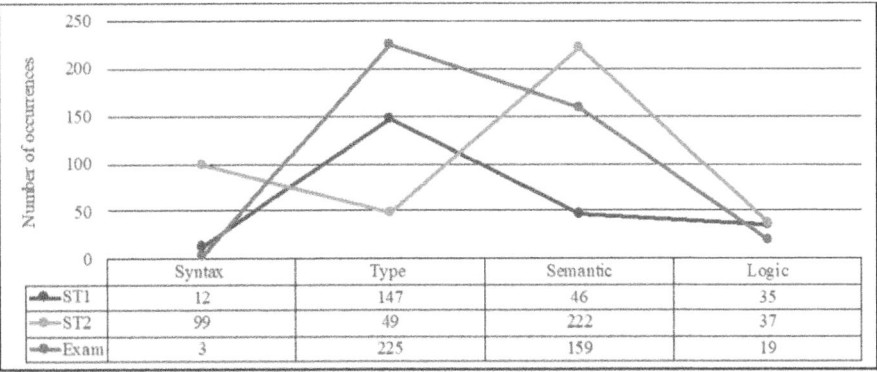

Fig. 4. Error categories for summative assessments

the number of occurrences for the four error categories in each of the three summative assessments.

It is interesting to note how the most common error category change between these assessments. While the semantic errors category was the most common in ST2 (with 222 occurrences), the type errors category was the most common in both ST1 (147 occurrences) and the Exam (225 occurrences). Since ST1 took place after Prac5, the prominence of type errors in ST1 correspond with the high frequency of type errors in the first five practical assignments (see Fig. 2). The same observation applies to ST2 where the high frequency in semantic errors from Prac7 and Prac8 (see Fig. 2) continued in this test. This could be an indication that at the time of ST1 and ST2, the novices still lacked the necessary knowledge to deal with the errors made in the directly preceding formative assessments. The overall prominence of type errors in the Exam, however, remains unexplained–especially since semantic errors were most prominent in the three assessments (Prac7, Prac8 and ST2) that directly preceded the examination. It should be noted that the examination covered all the content discussed during the academic term.

Summary. The findings as presented in this sub-section, provide valuable insights regarding the fluctuations in common programming errors (i.e. most common errors and most common error categories) made by novice programmers during an academic term. The changes in programming errors can be attributed to all the different new topics that novices have to learn within a short time, combined with their overall lack of basic programming knowledge [4, 13]. It is extremely important that instructors realize that the development process of novice programmers is not linear. Novices tend to re-encounter the same common errors multiple times [35] while new errors can also be introduced as new content is covered.

5 Conclusion and Recommendation

Novice programmers face various challenges when learning to write computer programs. Due to their limited programming knowledge, these novices are bound to make unnecessary errors that they often find difficult and time-consuming to correct. This study aimed to (1) identify the common programming errors made by a select group of novice programmers, and (2) analyze how these common errors changed at different stages during an academic term. Using a manual, structured content analysis of 684 programming artefacts, this study identified 21 common programming errors made by novices when they write code. The identified errors were classified into four categories: syntax, semantic, logic and type errors. The type error **A** (Invoking methods with wrong arguments) and semantic error **C** (Class claims to implement an interface but does not implement all the required methods) were the most frequent errors made by novices. With semantic errors and type errors also identified as the most frequently occurring error categories, these are the errors that instructors need to particularly focus on as part of their teaching and learning strategies when teaching novice programmers.

Furthermore, as part of an efficient teaching and learning strategy for programming, it is crucial that instructors monitor novices' progress throughout the entire semester. Monitoring of novices' learning processes (including the errors they most commonly make) could help to identify individuals who are at risk of failing the programming course. Consequently, instructors will be in a better position to provide targeted help and proper intervention to address concepts that novices find difficult to master. If novices are made aware of these common errors more explicitly, it could also help them to avoid such errors which could potentially reduce their programming struggles and frustrations as well as improve their chances of passing the programming course. Knowledge of the identified common errors and error categories could assist programming instructors in making informed decisions while adapting the teaching and learning approaches they utilize for the teaching of novice programmers. Such interventions could also form the basis of a longitudinal study evaluating the changes in students' code writing errors in response to educational interventions over multiple semesters.

References

1. Ahadi, A., Lister, R., Lal, S., Hellas, A.: Learning programming, syntax errors and institution-specific factors. In: Proceedings of the 20th Australasian Computing Education Conference, pp. 90–96. ACM, New York (2018). https://doi.org/10.1145/3160489.3160490

2. Albrecht, E., Grabowski, J.: Sometimes it's just sloppiness-studying students' programming errors and misconceptions. In: Proceedings of the 51st ACM Technical Symposium on Computer Science Education, pp. 340–345. ACM, New York (2020). https://doi.org/10.1145/3328778.3366862
3. Alqadi, B.S., Maletic, J.I.: An empirical study of debugging patterns among novice programmers. In: Proceedings of the 2017 ACM SIGCSE Technical Symposium on Computer Science Education, pp. 15–20. ACM, New York (2017). https://doi.org/10.1145/3017680.3017761
4. Altadmri, A., Brown, N.C.C.: 37 million compilations: investigating novice programming mistakes in large-scale student data. In: Proceedings of the 46th ACM Technical Symposium on Computer Science Education, pp. 522–527. ACM, New York (2015). https://doi.org/10.1145/2676723.2677258
5. Altadmri, A., Kölling, M., Brown, N.C.C.: The cost of syntax and how to avoid it: text versus frame-based editing. In: Proceedings of the 2016 IEEE 40th Annual Computer Software and Applications Conference (COMPSAC), pp. 748–753. IEEE (2016). https://doi.org/10.1109/COMPSAC.2016.204
6. Alzahrani, N., Vahid, F., Edgcomb, A., Lysecky, R., Lysecky, S.: An analysis of common errors leading to excessive student struggle on homework problems in an introductory programming course. In: Proceedings of ASEE Annual Conference. ASEE, Riverside (2018)
7. Becker, B.A., et al.: Unexpected tokens: a review of programming error messages and design guidelines for the future. In: Proceedings of the 2019 ACM Conference on Innovation and Technology in Computer Science Education, pp. 253–254. ACM, New York (2019). https://doi.org/10.1145/3304221.3325539
8. Becker, B.A., Mooney, C.: Categorizing compiler error messages with principal component analysis. In: Proceedings of the 12th China-Europe International Symposium on Software Engineering Education (CEISEE 2016), pp. 28–29. CEISEE, Shenyang (2016)
9. Becker, B.A.: An effective approach to enhancing compiler error messages. In: Proceedings of the 47th ACM Technical Symposium on Computing Science Education, pp. 126–131. ACM, New York (2016). https://doi.org/10.1145/2839509.2844584
10. Brass, S., Goldberg, C.: Semantic errors in SQL queries: a quite complete list. J. Syst. Softw. **79**(5), 630–644 (2006). https://doi.org/10.1016/j.jss.2005.06.028
11. Bosse, Y., Redmiles, D., Gerosa, M.A.: Pedagogical content for professors of introductory programming courses. In: Proceedings of the 2019 ACM Conference on Innovation and Technology in Computer Science Education, pp. 429–435. ACM, New York (2019). https://doi.org/10.1145/3304221.3319776
12. Brown, N.C., Altadmri, A.: Investigating novice programming mistakes: Educator beliefs vs. student data. In: Proceedings of the 10th Annual Conference on International Computing Education Research, pp. 43–50. ACM, New York (2014). https://doi.org/10.1145/2632320.2632343
13. Canedo, E.D., Santos, G.A., Leite, L.L.: An assessment of the teaching-learning methodologies used in the introductory programming courses at a Brazilian university. Inform. Educ. **17**(1), 45–59 (2018). https://doi.org/10.15388/infedu.2018.03
14. Denny, P., Luxton-Reilly, A., Carpenter, D.: Enhancing syntax error messages appears ineffectual. In: Proceedings of the 2014 Conference on Innovation and Technology In Computer Science Education, pp. 273–278. ACM, New York (2014). https://doi.org/10.1145/2591708.2591748
15. Denny, P., Luxton, R.A., Tempero, E.: All syntax errors are not equal. In: Proceedings of the 17th ACM Annual Conference on Innovation and Technology in Computer Science Education, pp. 75–80. ACM, New York (2012). https://doi.org/10.1145/2325296.2325318
16. Ettles, A., Luxton-Reilly, A., Denny, P.: Common logic errors made by novice programmers. In: Proceedings of the 20th Australasian Computing Education Conference, pp. 83–89. ACM, New York (2018). https://doi.org/10.1145/3160489.3160493

17. Farrokhi, F., Mahmoudi-Hamidabad, A.: Rethinking convenience sampling: defining quality criteria. Theory Pract. Lang. Stud. **2**(4), 784–792 (2012). https://doi.org/10.4304/tpls.2.4.784-792
18. Figueiredo, J., Lopes, N., García-Peñalvo, F.J.: Predicting student failure in an introductory programming course with multiple back-propagation. In: Proceedings of the 7th International Conference on Technological Ecosystems for Enhancing Multiculturality, pp. 44–49. ACM, New York (2019). https://doi.org/10.1145/3362789.3362925
19. Hristova, M., Misra, A., Rutter, M., Mercuri, R.: Identifying and correcting Java programming errors for introductory computer science students. ACM SIGCSE Bull. **35**(1), 153–156 (2003). https://doi.org/10.1145/611892.611956
20. Johnson, W.L.: Understanding and debugging novice programs. Artif. Intell. **42**(1), 51–97 (1990). https://doi.org/10.1016/0004-3702(90)90094-G
21. Jackson, J., Cobb, M., Carver, C.: Identifying top Java errors for novice programmers. In: Proceedings Frontiers in Education 35th Annual Conference, p. T4C. IEEE, Indianapolis (2005). https://doi.org/10.1109/FIE.2005.1611967
22. Kazeem, O.N., Abiola, O.A., Akinola, S.O.: Bug pattern analysis of codes produced by beginner programmers using association rule mining technique. Univ. Ibadan J. Sci. Logics ICT Res. **4**(1), 10–24 (2020)
23. Koerber, A., McMichael, L.: Qualitative sampling methods: a primer for technical communicators. J. Bus. Tech. Commun. **22**(4), 454–473 (2008). https://doi.org/10.1177/1050651908320362
24. Lakkaraju, H., et al.: A machine learning framework to identify students at risk of adverse academic outcomes. In: Proceedings of the 21st ACM SIGKDD International Conference on Knowledge Discovery and Data Mining, pp. 1909–1918. ACM, New York (2015). https://doi.org/10.1145/2783258.2788620
25. Malik, S.I., Mathew, R., Hammood, M.M.: PROBSOL: a web-based application to develop problem-solving skills in introductory programming. In: Al-Masri, A., Curran, K. (eds.) Smart Technologies and Innovation for a Sustainable Future. ASTI, pp. 295–302. Springer, Cham (2019). https://doi.org/10.1007/978-3-030-01659-3_34
26. Malik, S.I., Mathew, R., Al-Nuaimi, R., Al-Sideiri, A., Coldwell-Neilson, J.: Learning problem solving skills: comparison of E-learning and M-learning in an introductory programming course. Educ. Inf. Technol. **24**(5), 2779–2796 (2019). https://doi.org/10.1007/s10639-019-09896-1
27. McCall, D., Kölling, M.: A new look at novice programmer errors. ACM Trans. Comput. Educ. **19**(4), 1–30 (2019). https://doi.org/10.1145/3335814
28. McCall, D., Kölling, M.: Meaningful categorisation of novice programmer errors. In: 2014 IEEE Frontiers in Education Conference (FIE) Proceedings, pp. 1–8. IEEE (2014). https://doi.org/10.1109/FIE.2014.7044420
29. Plonka, L., Sharp, H., Van der Linden, J., Dittrich, Y.: Knowledge transfer in pair programming: an in-depth analysis. Int. J. Hum. Comput. Stud. **73**, 66–78 (2015). https://doi.org/10.1016/j.ijhcs.2014.09.001
30. Plowright, D.: Using Mixed Methods: Frameworks for an Integrated Methodology. SAGE, London (Kindle edition) (2011)
31. Qian, Y., Lehman, J.: An investigation of high school students' errors in introductory programming: a data-driven approach. J. Educ. Comput. Res. **58**(5), 919–945 (2020). https://doi.org/10.1177/0735633119887508
32. Rahman, M., Watanobe, Y., Nakamura, K.: Source code assessment and classification based on estimated error probability using attentive LSTM language model and its application in programming education. Appl. Sci. **10**(8), 2973 (2020). https://doi.org/10.3390/app10082973

33. Robins, A.V.: Novice programmers and introductory programming. In: Fincher, S., Robins, A. (eds.) The Cambridge Handbook of Computing Education Research, pp. 327–376. Cambridge University Press, Cambridge (2019)
34. Schliep, P.A.: Usability of error messages for introductory students. Sch. Horiz. Univ. Minn. Morris Undergrad. J. **2**(2), 5–11 (2015)
35. Smith, R., Rixner, S.: The error landscape: characterizing the mistakes of novice programmers. In: Proceedings of the 50th ACM Technical Symposium on Computer Science Education, pp. 538–544. ACM, New York (2019). https://doi.org/10.1145/3287324.3287394
36. Steenmeijer, L.D.: Increasing the consistency of feedback on programming code by using a tagging system (Bachelor's thesis). University of Twente, Twente (2020). http://essay.utwente.nl/82198/
37. Tabanao, E.S., Rodrigo, M.M.T., Jadud, M.C.: Identifying at-risk novice java programmers through the analysis of online protocols. In: Proceedings of Philippine Computing Science Congress, pp. 1–8. Archīum Ateneo (2008)
38. Türkmen, G., Caner, S.: The investigation of novice programmers' debugging behaviors to inform intelligent e-learning environments: a case study. Turkish Online J. Dist. Educ. **21**(3), 142–155 (2020). https://doi.org/10.17718/tojde.762039
39. Tuugalei, I., Mow, I.C.: Analyses of student programming errors in java programming courses. J. Emerg. Trends Comput. Inf. Sci. **3**(5), 739–749 (2012)
40. Zehetmeier, D., Böttcher, A., Brüggemann-Klein, A., Thurner, V.: Development of a classification scheme for errors observed in the process of computer programming education. In: Proceedings of the 1st International Conference on Higher Education Advances (Head 2015), pp. 475–484. Universitat Politècnica de València, València (2015). https://doi.org/10.4995/HEAd15.2015.356

Ten Years in the Trenches of a Doubtful Science: An Autoethnographic Investigation of Five Challenges of Teaching in Information Systems

Daniel B. le Roux

Information Science, Stellenbosch University, Stellenbosch, South Africa
dbleroux@sun.ac.za

Abstract. The field of Information Systems (IS) is characterised by a number of properties that combine to create particular challenges for teaching and learning. These properties include its close ties with rapidly advancing digital technology, its interdisciplinary nature and its general lack of a strong, broadly agreed-upon theoretical core. In the present study, I undertake an autoethnographic investigation of five key challenges for teaching in IS which result from one or more of these properties. To provide theoretical scaffolding for the investigation I adopt Weick's theory of sensemaking and apply it to investigate both my own and my students' processes of sensemaking. I propose that to effectively navigate these challenges and prepare students for careers in IS, lecturers should aim to develop their students' technical and social skills sets such that they are able to navigate the uncertainty and ambiguity that characterise socio-technical systems in practice. To this end, I outline the strategies I have adopted in my attempts to achieve this aim.

Keywords: Information systems · Teaching · Challenges · Autoethnographic · Sensemaking

1 Introduction

Since its emergence in the early 1960s, the field of Information Systems (IS) has rapidly developed and diversified as scholars have attempted to keep pace with the advancement of digital technology and its diffusion across all business and social spheres [6, p. 136]. The range of topics that potentially falls within the scope of IS has expanded so fast that the determination of the field's focus and boundaries are matters of ongoing debate among its members. This dynamism makes IS an exciting domain for teaching and research and, as industry demand for IS skills have increased, programme enrolments have grown [15]. In addition to fast-paced change, the field has, from its inception, been characterised by interdisciplinarity, recognising the important principle that information technologies do not function in isolation and that the boundary between the information system and its setting (e.g., societal, organisational or other) is often difficult to determine.

Despite its broad recognition as an academic discipline today, the development of IS has been (and still is) marred by a range of challenges. In its early years, scepticism about its legitimacy and positioning were expressed by leading scholars. Dearden, in 1972, argued that IS is a mirage (referenced in King and Lyytinen [20, p. 540]) and questioned the fundamental principles underlying the field's emergence. Kling, in 1980, described the field as "an arena yearning to be a discipline" (referenced in King and Lyytinen [20, p. 540]) while Keen contested that the field lacked a theoretical base (referenced in Culnan [12, p. 169]). Mum- ford and colleagues, in 1985, described IS as "a doubtful science" characterised by poor intellectual and methodical rooting [20, p. 540]. Questions about the field's legitimacy and distinctiveness continued to be raised in the 1990s with Avgerou et al. [6] describing it as ill-defined and facing problems in terms of recognition and legitimacy. IS has, notwithstanding these challenges, continued to grow and expand but as information technologies infiltrate every corner of human life these concerns and debates are likely to continuously characterise the fields in which they are studied.

Against this backdrop, an interesting and important tension that is often overlooked is that IS lecturers at universities across the globe have to design curricula and teach courses despite the ever-shifting thematic landscape and theoretical uncertainty of their field. This presents a significant challenge for both lecturers and students. To teach effectively, lecturers are required to continuously monitor the rapid developments in the field and redesign their courses at short intervals. Students, on the other hand, have to come to terms with the transitory nature of technological skills and best practices in IS. Doubtless, IS is not the only field that presents this challenge, but the rapidity of digital technology advancement and its uptake across industries and society make IS a particularly interesting and important case. While a fairly large collection of literature addresses the theoretical and conceptual ambiguity which result from these dynamic forces, fewer studies have considered how lecturers manage the resulting challenges in the course of their day-to-day teaching responsibilities.

In the present paper, I aim to address this gap by reporting an autoethnographic investigation of my role as a lecturer in IS over the past decade. To achieve this I adopt Weick's theory of sensemaking [33–36] as a theoretical lens to systematically analyse five key challenges I face as an IS lecturer which result (directly or indirectly) from the dynamic, interdisciplinary and theoretically indistinct nature of the field. For each challenge, I outline the factors which exasperate it and I provide brief descriptions of the mindset and strategies which I adopt to manage or cope with it. The scope of the paper is demarcated in two imported ways. Firstly, the emphasis falls primarily on the teaching dimension of academic work. The task portfolios of academics obviously include a much broader range of responsibilities but, while acknowledging these, they are not of interest in this investigation. Secondly, while most of my students are undergraduates that have little or no work experience, I also teach to post-graduate students that study on a part-time basis. The difference between these two student groups is important and requires thoughtful adaptation of curriculum and teaching approaches. While I briefly address these differences, the emphasis falls primarily on the challenges of teaching to the former group—full-time undergraduate students without work experience.

2 Method

2.1 Autoethnography as Method

"Autoethnography is a research method that uses personal experience ('auto') to describe and interpret ('graphy') cultural texts, experiences, beliefs, and practices ('ethno')" [2, p. 1]. The method adopts the principle that personal experience is infused with the norms and expectations of a particular setting and enables the researcher to utilise accounts of their personal experiences to "complement or fill gaps in, existing research" [2, p. 1]. Many scholars have adopted the method as a means to critique (often problematic) ethnographic accounts by researchers without personal experience of the culture under investigation (i.e., "cultural outsiders") [2, p. 4]. The method, accordingly, is sensitive to the manner in which the findings of social studies are "inextricably tied to the vocabularies and paradigms the scientists used to represent them" [13, p. 274].

Importantly, the method places substantial emphasis on personal experience, enabling research to describe and analyse "moments of everyday experience which cannot be captured through more traditional research methods" [2, p. 4]. Central to this retrospective and selective investigation of personal experience is the reporting of epiphanies which need to be analysed by contrasting personal experience against existing research [13, p. 276]. Epistemologically, autoethnography adopts the premise that "a social scientist who has lived through an experience and has consumed, unanswered questions about it can use introspection as a data source and, following accepted practices of field research, study him- or herself as with any 'n' of 1" [31, p. 148]. While emphasis falls on experience and introspection, autoethnographers, as part of the research process, analyse data sources like personal notes, records, transcripts and other artefacts. It is accepted the interrogation of these personal records is often "intensely personal" [31, p. 150], the aim being to facilitate reflection about and conveying a "patchwork of feelings, experiences, emotions, and behaviours that portray a more complete view of… life" [23, p. 10].

2.2 Autoethnography Studies of Teaching in Higher Education

Autoethnography is well-suited to investigations of the culture of work in higher education settings and has been used to this end with notable effect. For example, Hay et al. [19] review the substantial impact of Iain Hay's autoethnographic expression of his "core personal views on high-quality teaching" in Letter to a new university teacher [18]. The letter has been used internationally to guide the orientation of new lecturing staff. In it Hay "adopted an autoethnographic approach, presenting his thoughts in the form of a letter… to communicate with an audience he could envisage easily and that allowed personal commentary without seeming entirely self-centred" [19, p. 579]. Etherington [14], in a comparable article, provide suggestions on teaching practices for new lecturers. He acknowledges, for example, that upon commencement of his academic career "the very different roles of a researcher and a lecturer… were simply not obvious to me as a researcher who had just become a lecturer" [14, p. 1]. Leach [21] investigates her lived experience as an early-career academic through autoethnographic poetry, while Wilkinson [37] addresses the important issue of impostor syndrome in academe through an autoethnographic study of his role as an early career lecturer. [16] adopt an autoethnographic approach to explore the effects of audit culture and neoliberalism on university teaching, with a focus on the various demands that this culture places on lecturers. Lastly, Lourens [22], in a striking autoethnographic study, investigates the experiences of disabled employees within the academe in South Africa.

2.3 Research Process

The process adopted in the present investigation commenced from the position that teaching in IS is characterised by various challenges which can be attributed to three of the field's key properties: Its interdisciplinary nature, its close ties to rapidly advancing digital technologies and, finally, its lack of a broadly agreed-upon conceptual and theoretical core. I commenced with an introspective investigation of my role as lecturer and developed a list of the major challenges I faced, focusing specifically on challenges relating to, firstly, curriculum and course design (what I teach) and, secondly, pedagogy (how I teach). Throughout this process, I consulted a range of data sources including my personal collection of course outlines, lecture plans, reading lists and assessments compiled over the past 10 years. Of particular interest in this phase of the study were the curriculum and pedagogical changes I made year on year. These changes provided a useful lens into my attempts to improve the quality of my teaching, often by specifically addressing challenges or poor outcomes experienced in the previous year.

After compiling the initial list, it was refined by grouping overlapping challenges together and developing broader descriptions for each of the emerging categories to capture their essence. Each item in the finalised category list was then considered in relation to the three properties of IS as outlined above. In each instance, I considered, firstly, how the properties of IS impact (e.g., underlie, trigger, exasperate, ameliorate etc.) the challenge and, secondly, wrote descriptions of these relationships.

In the final phase of the investigation, I returned to the data sources and searched for evidence of attempts to address these challenges through adaptations to curriculum or pedagogy. In particular, I focused on the manner in which I changed my teaching of the same topic or theme from one academic year to the next and reflected on the motivations for and effectiveness of these changes. This enabled me to describe the interaction between refinement in my personal knowledge of the field and the manner in which I communicated to my students.

3 Theoretical Points of Departure

To support the conceptual and theoretical framing of the research I adopt Weick's theory of sensemaking [33–36] which describes the processes by which humans form meaningful understandings of reality based on their ongoing streams of experience. Central to the theory is the principle that that human actors perceive reality as a constant stream of experiences through ongoing processes of "automatic information processing" [35, p. 14]. As part of these processes, we use our senses to extract "present moments of experience" (termed cues) from the environment and place them in "perceptual frameworks" (or frames) to form plausible understandings of reality [35, p. 109–111]. Meaning, consequently, is the product of a person's ability to construct a relationship between a cue and a frame [35, p. 109–111]. Over time we develop a larger and increasingly meaningful frame repertoire, enabling us to create meaning of a greater variety of cues.

Weick describes seven properties of the sensemaking process, two of which are of particular relevance in this paper. Firstly, sensemaking is ongoing. Our streams of experience have no objectively definable starting or ending points but, to cope with this constant flow, we bracket particular parts of these streams and construct distinguishable moments from them [35, p. 43]. This enables people to assign absolute categories to reality, but also to ignore large parts of it [35, p. 44]. The second relevant property of sensemaking is that it is always retrospective. Any experience, Wieck argues, is only identifiable when a person steps outside the ongoing processes of automatic information processing and directs attention retrospectively. Meaning, consequently, is produced by paying attention retrospectively to particular brackets of experience or cues.

In addition to the properties of sensemaking, Weick describes various forms that the substance of sensemaking can take—the things people "draw on" when they make sense [35, p. 109]. Of particular relevance in the present study are theories of action which are symbolic representations of what action, under what circumstances, should be used to achieve some desired outcome [35, p. 122]. Theories of actions provide a useful conceptual tool to describe the frame repertoire one develops through experience in a particular occupation. Weick warns, however, that such theories are necessarily abstractions and that slippage between theory and action in practice will exist.

Three levels of sensemaking are of interest in this paper. Firstly, and primarily, the paper concerns my personal, subjective sensemaking processes over the past decade. This includes, on one level, my sensemaking of IS as an academic field but it also includes, on a second level, my sensemaking of the learning processes of my students. Central, in this regard, is my growing appreciation of the various learning challenges which students of IS face, and how these can be managed through different interventions. My personal theories of action, accordingly, are the products of the retrospective sensemaking of experiences I have had during the implementation of these interventions. The interplay between these two levels is also important—as my frames of IS have become more meaningful, my theories of action have evolved. The third level of interest is the sensemaking processes undertaken by students during learning. As a lecturer, my view of these processes is defined by my various forms of interaction with students. This includes class, lab and personal conversations and discussions, but it also includes assessment of their work which reflects the quality of their frames of the course content.

4 Challenges

In the sections which follow I describe each of the challenges identified, their associations with the properties of IS and, finally, provide an outline of the curriculum and pedagogical strategies I have adopted to manage them. The challenges are presented in an order which reflects their interdependence such that the management of earlier challenges often forms important building blocks for the successful management of those following upon them.

4.1 Challenge 1: Navigating Conceptual Ambiguity

After a decade of teaching and research in IS I've come to appreciate (even enjoy) the conceptual intricacies of my field. Over time my frames of the key constructs in the discipline have become more nuanced, reflecting the knowledge that their meanings are context-dependent and, as such, resist single, unambiguous definitions. However, as a lecturer, I see, year on year, how confusing this conceptual "briar patch" [11] can be to my students. In my experience there are very few IS undergraduate students who, even in their final year of study, have an unambiguous understanding of the key constructs in the field. This, I believe, is not due to any shortcomings of their own or of their lecturers. It is, rather, due to the nature of the concepts themselves.

Personally, a deeper appreciation of this challenge emerged during my post-graduate research projects when I learned that the task of specifying definitions for even the most basic concepts in the field (e.g., data, information and knowledge) is thwarted with a range of historical, logical and technical pitfalls and obstructions (see e.g. [1, 7, 8, 11, 27, 32, 38]). This contrasts the field to more established disciplines, particularly those in the so-called hard sciences where concepts tend to have exact meanings. Ambiguity about the meaning of first principles has an obvious knock-on effect on all the higher-level concepts in which they are utilised. Not surprisingly, as pointed out by Carvalho [10], the information system concept is used in literature to denote the number of fundamentally different phenomena.

I have found that the challenge of conceptual ambiguity can impact teaching quality in various ways. My most significant challenge in this regard has been the degree to which higher levels of argumentation and reasoning become hindered by a failure to establish conceptual clarity at more basic levels. For example, in a course I taught some years ago I covered the various ways in which the determination of information system success can be approached. In the lead-up to this part of the course, I made a substantial effort to cultivate a socio-technical understanding of information systems among my students, highlighting the importance of achieving alignment between technical and non-technical system components. However, when we started to discuss the notion of information system success, my students would consistently default to a more narrow (i.e., technical) conceptualisation and, consequently, their ideas about success would be dominated by technical correctness and quality. Thus, the idea of the information system as a socio-technical phenomenon was accepted in a sort of ceremonial manner, but the important implications of this conceptual decision seemed to be forgotten or ignored at higher levels of reasoning.

In my early years of teaching I often tried to sidestep conceptual intricacies by encouraging students to keep their eyes on the real-world problems to be solved rather than fretting over accurate definitions for the concepts we use to describe them. IS practitioners, I would say, have no time to quibble over the meanings of terms, they need to solve problems fast and effectively. However, this approach is only partially helpful in a teaching context where unambiguous communication depends upon a shared lexicon. Moreover, it limits my ability to convey more nuanced ideas about systems operation like congruence, adaptation and system evolution. An alternative and perhaps more successful intervention has been to introduce fresh concepts (neologisms) into the course lexicon, rather than attempting to reframe students' existing, often tainted, understandings of concepts. This tabula rasa approach has two benefits. Firstly, it creates an occasion for sensemaking because students are confronted with a novel cue. Secondly, it offers me a degree of freedom from preconceived ideas about the field and its domains. One example of this has been to utilise the work systems concept (as opposed to information systems) and it's associated framework as proposed by Alter [3–5]. My students, at the outset of the course, have no existing frames for the notion of a working system. This enables me to instil in them a socio-technical conceptualisation upon which more nuanced layers of understanding can be based with confidence that there is adequate alignment between my own and their frames.

4.2 Challenge 2: Teaching for a Wide Range of Roles

Of all the questions students have asked me in the past decade, the one which I have received most frequently is: "What type of work will I do one day?" It is often asked by a student looking simultaneously confused and anxious, uncertain whether they really want to be on the career path they have chosen. I've come to appreciate that students' uncertainty about the nature of their future careers is both justified and difficult to address. Graduates from the programmes in which I teach go on to perform a wide range of professional roles, including, among others, software developers, system analysts, business analysts, data scientists, testers, project managers, IT managers and consultants. In addition to the list being rather long, differentiation between these roles have become increasingly difficult and transient—I've frequently had graduates become employed in roles I've never heard of before. It is my view that this dynamic nature of IS practice is to be embraced rather than avoided as it reflects an ongoing evolution of the organisational forms best suited to achieving its goals. Understandably, however, students are intimidated by what is perceived as a disorganised, ad-hocratic industry and disappointed when lecturers are unable to sketch a more precise picture of their future career possibilities. I have also found this to be an aspect of the field that makes it unattractive to prospective students who want to become something (e.g., a doctor, an engineer, an auditor etc.).

In my experience, there are three important factors that serve to contextualise this challenge in IS. Firstly, the continuous emergence of new types of professional positions or renaming of existing ones resembles the industry's efforts to find effective and efficient organisational configurations. Secondly, the nature of work in IS implies that it is often difficult to clearly demarcate roles and overlap in task portfolios is common. Thirdly, the culture in many tech-oriented organisations is characterised by relaxed or agnostic attitudes towards naming conventions for professional roles in which versatile individuals are employed. For example, in many smaller or start-up companies workers are told to pick their own job titles.

To guide students' sensemaking of the turbulence in IS practice, I often utilise a metaphor that was used in a panel session I attended at the 2009 International Conference on Information Systems (ICIS) in Phoenix. The panel discussed the contentious topic of theory development in IS and the various challenges which obstruct this endeavour. During the concluding remarks, one of the panellists made the comment that scholars and students in IS are required to develop sea legs—the ability to keep one's balance and not feel seasick when onboard a moving ship. I found the metaphor to be particularly striking and I've used it to encourage my students to become comfortable in an unstable environment where one is expected to continuously learn and develop. Rather than framing this as problematic or a source of anxiety, I urge them to view it as exciting and rewarding by contrasting IS professionals to workers who perform the same set of repetitive tasks year after year. Additionally, I encourage them to reframe feelings of impostor syndrome as opportunities for learning and growth and to utilise these as motivators to enhance their personal knowledge and skills repertoire.

A second strategy I've adopted to manage this challenge is to invite a career advisor to present a guest lecture at the end of a course I present to final-year students. The career advisor specialises in IS and IT occupations and has a strong understanding of the market trends. Moreover, she has, over the years, placed many graduates from the programmes I teach in and has good knowledge of their curricula. During the lecture she addresses a range of topics, exposing students to the current market trends and optimal strategies for finding suitable entry-level occupations. In my experience, this orientation exercise plays an important role in supporting students' efforts to make sense of the job market, and it gives them a degree of confidence in the knowledge they have obtained during their degree programmes.

4.3 Challenge 3: Demystifying the Social Issues in IS

As a lecturer, I have generally found it to be substantially less challenging to teach courses of which the outcomes concern technical skills or proficiency like computer programming or database design. These hard disciplines lend themselves well to traditional, reductionist modes of teaching and learning in which the lecturer, generally well-versed in the relevant technology, guides the student towards the correct or optimal solution to a problem by teaching particular techniques. Much more challenging, particularly at the undergraduate level, has been courses that address non-technical themes in IS. Undergraduate students without work experience do not have meaningful frames of organisational systems and culture, limiting the degree to which one can effectively communicate about the social aspects of information systems operation.

This challenge perhaps best represents the difference between full-time undergraduate (non-working) and part-time postgraduate students that have work experience. In my classes with post-graduate students, questions and discussions are dominated by a wide range of non-technical issues (e.g., budget allocation, project management, adoption, adaptation, workarounds, shadow systems, user training, manager buy-in etc.). These discussions suggest that their work experience has led to the development of frames of system success that are sensitive to a wide range of non-technical factors. I have tried to instil these same holistic frames among my undergraduate students by working through case studies and the results of research studies on system success factors. However, these efforts have not, in my experience, had a considerable impact and I have developed the view that one can only really make sense of the complexities of organisational dynamics through actual work experience.

To manage this challenge, I have adopted two general strategies. The first is to emphasise the role of congruence or fit between social and technical system components. To this end, I adopt various theories/frameworks which describe fit in information systems, both at the level of the individual user (e.g., Task-technology fit [17]), and the level of the organisation as a whole (e.g., package-organisation alignment [25, 26, 28]). These theories/frameworks provide particularly useful lenses for the investigation of case studies of system failure as they guide students to consider the system holistically and focus on the nature of the relationships between system components as opposed to the technical success of hardware or software artefacts.

My second strategy has been to explore examples or cases in which the students themselves are actors in a system. This enables me to draw on their existing frames of systems and bring awareness to their subjective experiences as users. This enables discussion of how these experiences influence their future use intentions and, ultimately, the functioning of the system as a whole. I also use this approach to address the distinction between voluntary and compulsory systems use and explore their experiences of both. In my experience students tend to tacitly adopt the view that all information systems in business settings are characterised by compulsory use and, as a result, underestimate the agency of users in system adoption, operation and success.

4.4 Challenge 4: Teaching Soft Skills

While some IS graduates pursue careers that predominantly involve technical or analytical work (e.g. programmers or database administrators), the majority of graduates from the programmes I teach take up occupations that require a combination of technical/analytical and soft skills. I use the term soft skills here to refer to a collection of abilities relating to one's interaction with others. The role of Information Technology Business Analyst (ITBA) is perhaps the best example of a career path in which a large portion of the task portfolio involves various forms of social engagement (e.g., requirements elicitation, joint-application development, structured walkthroughs, SCRUM meetings, user training etc.). While I have found the classroom and lab to be well-suited to the teaching of analytical and technical skills, I have struggled to develop meaningful strategies to develop my students' soft skills. Moreover, while many ITBA handbooks emphasise the important role of these skills, they rarely provide meaningful guidance for teaching or practising social skills.

A central obstruction in this regard is students' ideas about the possibility of improving their social skill set. At the outset of an ITBA course, I normally run a quick poll to ask students, firstly, whether they believe they have a strong or weak social skillset and, secondly, whether they believe that this is an area in which one can improve through learning and training. I have found that, in general, a majority of students believe that one's social skills is primarily a product of personality and, as such, cannot be changed substantially through training. This shared belief has two effects—firstly, students who perceive themselves to have weak social skills suddenly become concerned about their career choice and, secondly, students are discouraged to participate in any teaching interventions aimed at social skill improvement.

To manage this challenge, I try to expose students to various cues that disconfirm their beliefs and encourage new sensemaking. This includes, for example, discussing cases of prominent individuals like Warren Buffet who has frequently stated that the advancement of his social skills has been the most important aspect of his education [24]. Buffet attributes much of his success in this regard to a course presented by Dale Carnegie, the contents of which have been captured in Carnegie's widely admired book

How to win friends and influence people [9]. I encourage my students, particularly those who perceive themselves to be socially awkward, to read the book.

I have also developed two teaching strategies to address this challenge. The first is to provide students with a strong set of practical guidelines for different types of social interactions typically encountered by professionals like ITBAs. In this regard, a book by Zwiers [39] has been particularly useful. Additionally, I have developed a series of practical sessions in which students engage in various forms of role-playing. These sessions typically run for 50 min and involve active participation by myself and a group of four to eight students at a time. During such a session one group of students will be tasked to perform a particular engagement technique (e.g., a brainstorming session with the client), with another group of students performing the roles of clients. Throughout such sessions I provide guidance and advice and, upon completion, we review the session as a group. While the role-playing environment is obviously artificial, the sessions, when managed well, provide a valuable sandbox for students to test their abilities and learn about their strengths and weaknesses. It also enables me to identify students who need particular support or guidance. Unfortunately, the sessions seem to be less effective if I do not participate and, as such, the intervention does not scale well to large class groups when there is limited class time available.

4.5 Challenge 5: Bridging the Research-Teaching Gap

Upon my appointment as an early-career lecturer in 2010, I attended an orientation course for new academic staff at the institution. The course ran over three days and covered a wide range of topics. A central theme across many of these topics concerned the challenge of balancing the teaching, research and administrative workloads that characterise typical academic positions. Presenters suggested that one should aim to find alignment between one's teaching responsibilities and research interests, enabling some degree of overlap between these responsibilities. Intuitively, this made sense to me as it would ensure that my course content covers the most recent developments in the specific domain. In retrospect, however, this Utopian vision was naive and I have, for the most part, been unable to achieve a meaningful alignment between the content I teach and the research I perform. I do not believe that this reflects any form of failure on my part, but rather that it is an inevitable product of a maturing and increasingly focused personal research agenda.

While curriculum design in IS is typically characterised by frequent updating of content and course structures, the general components of a model IS programme have remained fairly stable over the past decade. The AIS (Association for Information Systems) published an IS model curriculum in 2010 [30] with an initial draft of an updated version appearing in August 2020 [29]. In the same 10-year period my personal research interests have become focused on very particular topics which, while certainly falling within the general scope of IS, are much more specialised than the topics I cover when I teach undergraduate classes. In my experience, reputable academic journals are unlikely to publish research that addresses topics at an introductory level, and my undergraduate students are equally unlikely to find value in lectures that cover very specific, specialised research problems.

Despite this apparent irreconcilability, it would be false to state that my research has had no impact on my teaching. Even when there exist substantial differences between the topics covered in either, I am aware of their interaction in my sensemaking. Experience gained through the research process, independent of the method followed or question asked, always provides new insights which stimulate sensemaking and enhance one's understanding of the field. This inevitably flows through to the way I teach and interact with students. Similarly, interaction with students often stimulates sensemaking through debate and argumentation, triggering the continuous updating of my frames of their learning processes.

In addition to acknowledging the underlying interplay between research and teaching, I have also tried to bridge this gap through two particular strategies. The first is to encourage my students to continuously look for and challenge assumptions when studying theories or frameworks in prescribed textbooks. For example, when presenting a theory to a class, I ask them to consider its potential shortcomings or provide examples of cases that may falsify it. By stimulating this critical outlook, I create an opportunity to briefly discuss recent research findings that relate to the theory, raising their awareness of the limitations of an abstract artefact like a theory. The second strategy is to commence a lecture with a brief overview of a recently published study that relates, at least partly, to the lecture content. This serves the purpose of alerting students to the degree to which handbooks or textbooks often present an idealised picture of information systems development and operation, and that the real world is messy and unpredictable. It also serves to pique their interest and make them curious about various aspects and implications of the study, laying the groundwork for sensemaking.

5 Discussion

Ongoing processes of sensemaking characterise academic work across all fields. In IS, however, the field's intertwinedness with rapidly advancing digital technologies, its interdisciplinary character and its theoretical indistinctness combine to create various teaching and learning challenges for both lecturers and students. In this paper I've adopted an autoethnographic approach to outline five of these challenges, utilising Weickean sensemaking as a theoretical lens.

In their totality, the five challenges are reflective of the manner in which the dynamic nature of IS resists the specification of fixed or rigid curricula. The field requires, rather, that its members continuously monitor the trends in digital technology development and adoption, and make sense of how these impact human behaviour and organisational functioning and performance. Beyond the field's broad but generally weak theoretical base, there are no universal laws that enable accurate predictions about the development, dissemination and uptake of technologies. Consequently, IS scholars often find themselves playing catch-up as digital media become embedded in every corner of our lives. On a certain level, the idea of lecturing in such a turbulent field is in itself questionable.

Despite the ever-presence of so many uncertainties, I believe that IS lecturers can conduct their teaching in a manner that, on the one hand, embraces the field's dynamism and, on the other, equips students with the appropriate attitudes and skillsets to be successful professionals. This requires careful navigation of various challenges, some of which are described in this paper. How lecturers approach these challenges would depend upon their personal backgrounds and contexts, but the strategies proposed may serve to guide or inspire these efforts.

I have, over the past decade, become increasingly aware of my students' perspective of the challenges outlined here. Choosing a career path characterised by so much turbulence and uncertainty is a brave step, even if most of them do not fully appreciate this at the outset of their studies. Over time I have observed an evolution in my approach from teaching particular frameworks, techniques and skills, to creating learning experiences that cultivate sea legs. This not only requires that students become exposed to the non-deterministic nature of socio-technical systems in real-world settings, but that they develop confidence in their ability to navigate the resulting uncertainty. These learning experiences are not always comfortable for them, and I often receive complaints and requests for more specific instructions and guidance. However, from a sensemaking perspective, these experiences of uncertainty and ambiguity are valuable occasions for broadening frames of reference and developing increasingly nuanced appreciations of socio-technical systems.

Finally, as is the case in any autoethnographic study, the obvious limitations associated with such a strong reliance on personal experience should be acknowledged. The degree to which the challenges I outline are representative of challenges experienced by IS lecturers, in general, should be questioned. While I aimed to present them in an abstract manner, it might well be that they poorly reflect the experiences of my colleagues at other institutions.

References

1. Ackoff, R.: From data to wisdom. J. Appl. Syst. Anal. **16**, 3–9 (1989)
2. Moriarty, J.: Autoethnography. In: Analytical Autoethnodrama. BVER, pp. 37–52. SensePublishers, Rotterdam (2014). https://doi.org/10.1007/978-94-6209-890-9_4
3. Alter, S.: Defining information systems as work systems: implications for the IS field. Eur. J. Inf. Syst. **17**(5), 448–469 (2008)
4. Alter, S.: Work system theory: overview of core concepts, extensions, and challenges for the future. J. Assoc. Inf. Syst. **14**(2), 72–121 (2013)
5. Alter, S.: Sociotechnical systems through a work system lens: a possible path for reconciling system conceptualizations, business realities, and humanist values in IS development. In: Conference on Advanced Information System Engineering, pp. 1–15 (2015)
6. Avgerou, C., Siemer, J., Bjørn Andersen, N.: The academic field of information systems in Europe. Eur. J. Inf. Syst. **8**(2), 136–153 (1999)
7. Bates, M.J.: Fundamental forms of information. J. Am. Soc. Inf. Sci. Technol. **57**(8), 1033–1045 (2006)
8. Bawden, D.: Organised complexity, meaning and understanding: an approach to a unified view of information for information science. Aslib Proc. **59**(4/5), 307–327 (2007)
9. Carnegie, D.: How to Win Friends and Influence People. Simon and Schuster, New York (1936)
10. Carvalho, J.A.: Information system? Which one do you mean? In: Falkenberg, E.D., Lyytinen, K., Verrijn-Stuart, A.A. (eds.) Information System Concepts: An Integrated Discipline Emerging. ITIFIP, vol. 36, pp. 259–277. Springer, Boston (2000). https://doi.org/10.1007/978-0-387-35500-9_22

11. Checkland, P., Holwell, S.: Information, Systems and Information Systems, Making Sense of the Field. Wiley, Chichester (1998)
12. Culnan, M.J.: The intellectual development of management information systems, 1972–1982: a co-citation analysis. Manage. Sci. **32**(2), 156–172 (1986)
13. Ellis, C., Adams, T.E., Bochner, A.P.: Autoethnography: an overview. Historical Social Research/Historische Sozialforschung **36**(4 (138)), 273–290 (2011). https://www.jstor.org/stable/23032294
14. Etherington, T.R.: Seven simple suggestions to be a better teacher sooner: experiences of a nearly new lecturer. Preprint, PeerJ Preprints, September 2018. https://peerj.com/preprints/27185v1
15. Fitzgerald, B., Adam, F.: The status of the IS field: historical perspective and practical orientation. Inf. Res. **5**(4), 1–17 (2000)
16. Gonzalez-Calvo, G., Arias-Carballal, M.: Effects from audit culture and neoliberalism on university teaching: an autoethnographic perspective. Ethnogr. Educ. **13**(4), 413–427 (2018). https://doi.org/10.1080/17457823.2017.1347885
17. Goodhue, D.L., Thompson, R.L.: Task-technology fit and individual performance. Manag. Inf. Syst. **19**(2), 213–236 (1995)
18. Hay, I.: Letter to a new university teacher. HERDSA News **24**, 15–17 (2002)
19. Hay, I., Bartlett-Trafford, J., Chang, T.C., Kneale, P., Szili, G.: Advice and reflections for a university teacher beginning an academic career. J. Geogr. High. Educ. **37**(4), 578–594 (2013). https://doi.org/10.1080/03098265.2013.828276
20. King, J.L., Lyytinen, K.: Reach and Grasp. MIS Q. **28**(4), 539–551 (2004)
21. Leach, E.: The fractured "I": an autoethnographic account of a part-time doctoral student's experience with scholarly identity formation. Qual. Inq. **27**(3–4), 381–384 (2021). https://doi.org/10.1177/1077800420918895
22. Lourens, H.: Superscripting the academy: the difference narrative of a disabled academic. Disabil. Soc. 1–16 (2020). https://doi.org/10.1080/09687599.2020.1794798
23. Muncey, T.: Doing Autoethnography. Int. J. Qual. Methods **4**(1), 69–86 (2005). https://doi.org/10.1177/160940690500400105
24. Segal, G.Z.: Getting There: A Book of Mentors. Harry N. Abrams, New York (2015)
25. Sia, S.K., Soh, C.: An assessment of package–organisation misalignment: institutional and ontological structures. Eur. J. Inf. Syst. **16**(5), 568–583 (2007)
26. Soh, C., Sia, S.: An institutional perspective on sources of ERP package-organisation misalignments. J. Strateg. Inf. Syst. **13**(4), 375–397 (2004)
27. Stonier, T.: Information and the Internal Structure of the Universe: An Exploration into Information Physics. Springer-Verlag, London (1990). https://doi.org/10.1007/978-1-4471-3265-3
28. Strong, D.M., Volkoff, O.: Understanding organization-enterprise system fit: a path to theorizing the information technology artifact. MIS Q. **34**(4), 731–756 (2010)
29. The Joint ACM/AIS IS 2020 Task Force: IS2020 Competency Model for Undergraduate Degree Programs in Information Systems (Initial Draft) (2020). https://is2020.hosting2.acm.org/wp-content/uploads/2020/08/IS-2020-Draft-August-2020.pdf
30. Topi, H., et al.: IS 2010: Curriculum Guidelines for Undergraduate Degree Programs in Information Systems. Communications of the Association for Information Systems, vol. 26 (2010). https://aisel.aisnet.org/cais/vol26/iss1/18
31. Wall, S.: An autoethnography on learning about autoethnography. Int. J. Qual. Methods **5**(2), 146–160 (2006). https://doi.org/10.1177/160940690600500205
32. Weaver, W.: Some recent contributions to the mathematical theory of communication. In: The Mathematical Theory of Communication, pp. 1–16. University of Illinois Press, Urbana, Illinois (1949)

33. Weick, K.E.: The collapse of sensemaking in organizations: the Mann Gulch disaster. Adm. Sci. Q. **38**(4), 628 (1993)
34. Weick, K.: The non-traditional quality of organizational learning. Organ. Sci. **2**(1), 116–124 (1991)
35. Weick, K.: Sensemaking in Organizations. Sage, Thousand Oaks (1995)
36. Weick, K.: Improvisation as a mindset for organizational analysis. Organ. Sci. **9**(5), 543–555 (1998)
37. Wilkinson, C.: Imposter syndrome and the accidental academic: an autoethnographic account. Int. J. Acad. Dev. **25**(4), 363–374 (2020)
38. Zins, C.: Conceptual approaches for defining data, information, and knowledge. J. Am. Soc. Inform. Sci. Technol. **58**(4), 479–493 (2007)
39. Zwiers, V.: The Business Analyst: Information Technology's Paradigm Shift. Juta and Company Ltd., Cape Town (2013)

Mapping the Problem-Solving Strategies of Novice Programmers to Polya's Framework: SWOT Analysis as a Bottleneck Identification Tool

Pakiso J. Khomokhoana and Liezel Nel

Department of Computer Science and Informatics, University of the Free State,
Bloemfontein, South Africa
{khomokhoanap,nell}@ufs.ac.za

Abstract. The development of problem-solving skills continues to be a challenge in various disciplines including Computer Science. In this study, we used the principles of the Decoding the Disciplines (DtDs) paradigm to better understand the mental processes that novice programmers follow when answering source code comprehension (SCC) related questions. This understanding can be fundamental in helping novices to overcome problem-solving related challenges. While focusing on step 1 of the DtDs paradigm, the aim of this study was threefold. Firstly, we explored the problem-solving strategies utilised by novice programmers while they were attempting to answer SCC related questions. Secondly, the identified problem-solving strategies were mapped onto Polya's four problem-solving steps. Finally, we utilised a SWOT analysis as a tool to identify problem-solving related learning bottlenecks. This study utilised an integrated methodological approach where data was collected by means of asking questions, observations, and artefact analysis. Thematic analysis of the collected data revealed a range of problem-solving strategies that these novice programmers utilised while performing various SCC tasks. These strategies were then mapped onto Polya's problem-solving steps. Based on a SWOT analysis of these strategies, we identified six problem-solving bottlenecks that point to difficulties that are not sufficiently addressed in introductory CS courses.

Keywords: Decoding the disciplines · Problem-solving · Source code comprehension · Novice programmers · Computer Science education · Polya's framework · SWOT analysis

1 Introduction

High dropout rates in introductory computer programming courses remain a major concern for higher education educators across the world [1, 20, 22]. Computer programming is a complex and multi-faceted task as it requires not only conceptual and procedural knowledge but also skills to create, modify and comprehend computer code [13, 35, 38] in order to solve programming problems. Numerous studies [2, 19, 34] have identified a lack of problem-solving skills as one of the biggest challenges that novice programmers experienced. Another major challenge (which has been researched extensively) relates to how novice programmers comprehend (or interpret) pieces of source code [10, 18]. Lister et al. [18] regard the skill of source code comprehension (SCC) as a prerequisite to problem solving. While Conn and McLean [9] argue that novice programmers' lack of problem-solving skills stems from how problem solving is taught in schools, Belski [3] blames university educators for failing to properly develop this skill in students. This failure has been attributed to educators' tendency to focus on the teaching of syntactical and conceptual programming knowledge, and paying less (or no) attention to the strategic knowledge needed to appropriately and effectively solve programming problems [7, 19, 41]. This tendency is not exclusive to the Computer Science (CS) discipline. Middendorf and Shopkow [24] (p. 2) note that while educators (as practitioners) are familiar with the "approaches, techniques, and applications" of their specific academic disciplines, "they tend to organise their courses around specific contents rather than around the mental moves they want students to make".

The Decoding the Disciplines (DtDs) paradigm – formulated by Middendorf and Pace [23] – recognises that each discipline has its own unique ways of thinking that students need to master to succeed in their higher-level studies. As part of the initial step in the seven-step DtDs process, educators are encouraged to identify specific points where their students' learning are interrupted [11]. These points (referred to as bottlenecks) are likely to prevent students from mastering the basic disciplinary ways of thinking. Only once these bottlenecks have been identified can educators continue with the remaining DtDs steps of designing, implementing, and evaluating specific strategies to address the bottlenecks that their students are experiencing [26]. There are several examples of approaches that educators used to identify bottlenecks. These include educators' personal experiences in teaching a specific course [24, 28], problems discovered while grading student assignments [26], problems observed while watching students' attempts to solve a given problem [17], and clarification questions asked by students regarding assignment specifications [39]. A tool that could potentially also be used in this regard is a SWOT analysis [16]. Although mostly used in business environments, this strategic planning tool can help to analyse and position any situation in four regions: strengths, weaknesses, opportunities, and threats [36]. By analysing the mental moves novice programmers make while solving SCC problems, a SWOT analysis could potentially be used to, not only identify the problem-solving steps that students are performing well, but also the areas (potential bottlenecks) where students are not following the correct problem-solving steps.

The foundation of most modern problem-solving strategies (both general and in the field of CS) can be traced back to George Polya's [30] four basic problem-solving steps: understand the problem, devise a plan, carry out the plan, and evaluate the effectiveness of the plan. Since problem solving is a skill that is used daily by all human beings, it can be argued that the natural (unenhanced) problem-solving strategies followed by novice programmers should in theory show some resemblance to the original steps of Polya's framework. By mapping the problem-solving strategies followed by novice programmers while trying to answer an SCC question, it should be possible to identify shortcomings in their strategies. This, in turn, could point to potential learning bottlenecks that CS educators must specifically address.

This paper therefore attempts to answer the following three questions:

1. What are the problem-solving strategies utilised by novice programmers during SCC?
2. How do these strategies relate to Polya's four basic problem-solving steps?
3. How can a SWOT analysis of the SCC problem-solving strategies followed by novice programmers be used as a learning bottleneck identification tool?

In the remainder of this paper, a theoretical framework guiding this study together with a review of relevant background literature are presented in Sect. 2. This is followed by a discussion of the research design and methods in Sect. 3, and a presentation and interpretation of the results in Sect. 4. A SWOT analysis of the identified SCC strategies is presented in Sect. 5, followed by bottleneck identification in Sect. 6. Conclusions and recommendations for future research are presented in Sect. 7.

2 Theoretical Framework

Polya's [30] problem-solving framework provides the theoretical framework for this study. In the following sub-sections, the four basic steps of Polya's framework, together with examples of how these steps are typically executed by computer programmers during SCC tasks, are discussed in more detail.

2.1 Understand the Problem

It is impossible to solve a problem if the problem is not understood first. Polya [30] provides several questions that can be asked by the problem solver while trying to understand or comprehend a problem: Do you understand all the words used in the problem statement? What are you asked to find or show? Can you restate the problem in your own words? Can you think of a picture or diagram that might help you understand the problem? Is there enough information to enable you to find a solution?

As part of the understanding process, Chi et al. [8] suggest organising information around important concepts. As part of understanding a given problem, programmers typically apply strategies that include reading or re-reading problem requirements and/or related lines of code, and reasoning aloud [14, 25]. Other strategies such as highlighting or colouring some lines of code or text [31], writing comments [37], and making drawings or annotations (doodles) [18] are often utilised by programmers to gain a better understanding of the problem at hand. For SCC tasks, expert programmers will often scan the code from top to bottom to get a quick overview of the problem [40], while also making notes on important information. They can then easily refer to these notes (if needed) at a later stage [33].

2.2 Devise a Plan

Polya [30] suggests several of strategies for devising a problem-solving plan. These include guess and check, look for a pattern, make an orderly list, draw a picture, eliminate possibilities, solve a simpler problem, use symmetry, use a model, consider special cases, work backwards, use direct reasoning, use a formula, solve an equation, and be ingenious. In resonance with Polya [30], the Mathematics students in the study by Malloy and Jones [21] exhibited planning strategies that included drawing a picture or diagram, using patterns, making lists or charts, guessing and checking, working backward, using logical deduction, disregarding extra (unnecessary) data, as well as using multiple or a combination of strategies and logical deduction.

The problem-solving plan helps to arrange ideas together, which in turn helps a problem solver to gain an even better understanding of the problem [5]. With regard to SCC, Fitzgerald et al.'s [14] respondents recognised that one SCC question resembled other questions seen previously (pattern recognition) and, therefore, made assumptions based on previous answers. During the planning stage of SCC, programmers tend to identify possible test cases [25] to use in the execution of their plan. This helps them to avoid carrying out infinite tests. Similar to understanding a problem, programmers also make notes (annotations/doodles) on how their plan will be executed [18]. Ultimately, devising a plan will typically involve working through different scenarios and gauging the good as well as the bad points of each alternative [33]. During this process, a problem solver justifies all the decisions he/she arrives at [15]. This also enables problem solvers to remember important information that may be useful during the remainder of the problem-solving process [5].

2.3 Carry Out the Plan

When it comes to carrying out the plan, Polya [30] suggests that the chosen plan should be followed. If the plan does not work, it should be discarded, and another alternative plan should be selected. The decision to either discard or continue with the selected plan can only be made if the problem solver is continuously monitoring the plan – a task that requires metacognitive abilities [5]. Inherently, a lot of thinking and analysing are in operation during this problem-solving step [4]. In addition to applying surface thinking, the problem solver also needs to think critically and creatively while integrating prior knowledge as part of the process [12]. In carrying out a plan, the problem solver usually considers limited details that are only relevant to the selected plan [33]. Interestingly, Fitzgerald et al. [14] note that some of their participants would start from scratch with an SCC task whenever they were becoming confused.

2.4 Evaluate the Effectiveness of the Plan

To evaluate the effectiveness of the selected plan, Polya [30] suggests that the problem solver should take the time to reflect and look back at what has been done, what worked, and what did not work. This step emphasises the importance of assessing and possibly even re-assessing the identified solution to any programming or SCC related problem [12]. This will typically be an iterative process during which the programmer may discover even more knowledge about the problem and/or solution [19]. It is also not uncommon for programmers to document their reflections on a completed SCC task [37].

3 Research Methods

3.1 Design

The design of this study was narrative in nature and followed an integrated-methods research approach based on Plowright's [29] Frameworks for an Integrated Methodology (FraIM). Within this framework, both narrative and numeric data were collected by means of observations, asking questions, and artefact analysis. The study population consisted of final-year undergraduate CS students from a selected South African university. After participating in an SCC activity (as part of an earlier phase of this multi-phase research project), students who incorrectly answered the three most difficult questions (as determined during the earlier phase) were invited to take part in this phase of the study. The purposeful and convenient sample [27] consisted of the 10 students who agreed to partake in this part of the study. The purposefulness of the sample was based on the fact that the students had already completed four programming modules. However, they could still be regarded as novice programmers since they did not have any professional programming experience. Based on their underperformance in the mentioned SCC activity, it was highly likely that they were experiencing some bottlenecks in their learning. The sample was also convenient since the students were affiliated with the same department as the researchers.

3.2 Data Collection

The research activity consisted of individual sessions during which each participant had to use a think-aloud technique [32] to verbally explain his/her thinking processes while answering three selected SCC questions. These questions – C# versions of question 3 (Q3), question 6 (Q6) (see Appendix), and question 8 (Q8) from Lister et al.'s [18] study – were the same three questions that the selected participants were unable to answer correctly during an earlier phase of this research project. This data collection strategy can be regarded as a means of "asking questions" [29]. Time slots of 45 min were scheduled for each of the individual sessions. The first author (principal researcher) played the role of the interviewer by asking probing questions when required (i.e. in case of no progress or silence). Where deemed necessary, the interviewer also recorded some observations as an additional data collection strategy. The proceedings of each session were audio recorded with permission from the relevant participant.

3.3 Data Analysis

Since the participants had to verbalise their thoughts as part of the think-aloud process, the transcripts contained numerous illogical and repeated statements. We therefore cleansed the data, after which we familiarised ourselves with the data [6]. This was achieved by listening to the audio records numerous times as well as intensively and repeatedly reading the transcripts. This helped us to decide on a coding plan where the analysis would be guided by the data as it relates to our research questions. At this stage, we imported the 10 validated transcripts into NVivo 12 Professional for Microsoft Windows. After this, we developed the necessary codes (by creating several nodes) according to our research questions. We coded the data by highlighting and/or underlining text within the domain of the stated units of analysis. We then populated the created codes by moving the necessary text into them. We continuously revised the names of the codes and the relevant themes until recurrent themes were emerging. For each theme developed, the NVivo-generated frequencies of occurrence were used.

4 Results and Interpretation

The discussion in this section focuses on the problem-solving strategies identified from the data collected during the 10 think-aloud sessions. The discussion is grouped according to Polya's four problem-solving steps (see Sect. 2).

4.1 Understand the Problem

Analysis of both the think-aloud data and the interviewer's observations revealed that the participants utilised two main strategies to help them understand the problem.

Read Instructions and Identify Important Concepts or Terms. The typical first step in understanding the problem is to read the instructions or the problem statement [30]. This strategy was utilised by all the participants with a total of 35 occurrences observed. It is also key that the problem solver, while reading through the instructions, identifies important concepts and/or terms [8]. In the 75 occurrences of this strategy, the participants mentioned and/or underlined specific programming concepts and terms that they would have to understand in order to answer the given questions. These included Boolean values, parameters, initialisation, nested loops, arrays, array indexes, and the use of post- and pre-increment operators. By identifying all these concepts, participants were trying to form an initial understanding of crucial parts of the provided programming code.

Interpret the Problem Details. As part of understanding the problem [30], all the participants (29 occurrences) also tried to interpret more specific details from the provided SCC questions. As part of this interpretation process, they conducted a more thorough examination of specific pieces of source code and tried to explain the meaning thereof in an attempt to increase their level of understanding. At this stage, it also became apparent that the flawed interpretations made by some of the participants were likely to have a negative impact on their ability to ultimately solve the problem. Due to his misinterpretation of the combined index in the statement b[x[i]]= true; from Q3, P10 incorrectly concluded that "*the second* for *loop resets everything from the first* for *loop back to* true". By failing to recognise from the start that only the first three values in array b would be set to true, he was eventually unable to identify option B as the correct answer.

4.2 Devise a Plan

It was noted that none of the participants in this study attempted to formulate a definitive plan for solving the various SCC questions. They did, however, follow some of the strategies outlined in Sect. 2.2.

Eliminate Possibilities. After a quick look at the source code of Q6, P1 eliminated option C as the possible missing piece of source code. She said: "*Let me understand option C... if the index of an array at position* i *is bigger than the index of* i + 1*, I must return* false*, else return* b *which is already* false*. There is no way of returning* true*, so it is totally wrong, it can't be the missing source code*". This meant that she was left with only four options to consider, which simplified the remainder of the problem-solving process for her.

Make Notes. As part of devising a plan during the problem-solving process [30], problem solvers typically make various types of notes while they work through the problem [31, 37]. Seven participants utilised this strategy (with 24 occurrences). As part of his attempt to answer Q3, P5 made notes on the answer sheet to keep track of the values assigned to array b. While working through the second for loop he said: "*Set that value to* true*. So that is going to mean that 1, 2 and 3 is going to be* true". He then scratched out the three relevant false values in the notes he originally made when he was working through the first for loop. Although he did not make any further written notes, he continued to reference the written down values while he was working through the third for loop to determine the final value of the variable count. It should, however, be noted that note-making instances were also observed with the participants during the three other steps of the problem-solving process.

Use Test Cases. In devising a plan during problem solving, a problem solver can identify some test cases [25] to avoid conducting infinite tests. Four occurrences of this strategy were observed with three of the participants. As evidence that P6 used test cases, she said: "*Let me just eliminate this one for now, so that I don't waste time on irrelevant things ... [Interviewer: You have created a long array!] ... yeah it was quite long*". In formulating test cases for Q3, P2 decided to draw a trace table as part of his planning and reasoning. He later referred to this table when he had to determine his final answer, where he said: "*Since* i *is equal to 4, our value is 5. So, our answer here should be 5 according to this table*". P10 also created his own array and started adding values as he went through his reasoning for Q6: "*Well, I made my array and I just changed the values so that I can better understand what I was sorting out*". It is evident that these participants not only identified possible test cases, but specifically used these values during the execution of their problem-solving plans.

Work Backwards. While devising a plan, problem solvers can also work backwards to check or confirm some of their initial interpretations [30], and then take corrective actions if needed. Ten occurrences of this strategy were identified with six of the participants. While attempting to interpret the second for loop from Q3, P6 exhibited several signs of confirming initial understanding: "*... meaning* zero *is less than 5, yeah 5, wait... 5 ... yeah, it's 5... I think my answer is going to be B, wait...* x[i] *is 1, then 2... okay* *wait that's* `zero`, *that's 1. So now let's see,* `i` *is 0 ... eish, what do we call this step? Initialising? Yeah, I think so ... [Interviewer: Which one?] ... the one with* `b[x[i]]` *... [Interviewer: You are assigning another value.], Yeah, assigning it to* `true`". It is evident from this excerpt that while P6 attempted to confirm his initial understanding, he also resorted to seeking confirmation or clarification from the interviewer. He specifically wanted to confirm whether he was using the correct terminology while attempting to interpret the statement `b[x[i]]= true;` which is an example of an assignment and not an initialisation statement. It can therefore be deduced that although he was attempting to link whatever he was doing to known disciplinary concepts [8], there were still some gaps in his knowledge of basic programming terminology.

Pattern Recognition. Another strategy that problem solvers often utilise during the devising-a-plan step is pattern recognition. During SCC, this happens when the problem solver recognises similarities between different pieces of source code that allow for code sections to be considered collectively [14]. The pattern recognition strategy was observed with eight participants (20 occurrences). While looking at the initialisation of array `x` in Q3, P8 said: "*I am looking at the pattern for this, what was declared of* `x` *... so now by judging from the pattern that I see here, it's 1, 2 and then it becomes constant*". By "constant", P8 was referring to the value of 3 that was assigned to the last three elements of the `x` array. P6 also noticed a pattern in the headers of the first two `for` loops in Q3: "*The second* `for` *loop is still the same as the first one, but instead of taking the* `b` *length we take the* `x` *length, so all these are the same*". While P2 recognised the same pattern as P6, he also noted that there were actually similarities between the headers of all three `for` loops in Q3: "*All these loop statements look a little bit the same because all of them start at* `i = 0` *... and* `i++`, *so they are all correct here*". Identification of a pattern or range of patterns [21, 30] helps to reduce cognitive load as the problem solver is left with fewer unique possibilities to concentrate on [5].

4.3 Carry Out the Plan

In carrying out their plans, participants corrected their earlier misinterpretations and acknowledged areas where they experienced difficulties.

Correct Earlier Misinterpretations. In carrying out a plan, problem solvers can correct the misinterpretations they might have made in the previous problem-solving steps. This strategy was observed with three participants (five occurrences). After reviewing the pencil notes she made earlier in the problem-solving process, P9 realised that she had labelled the array indexes incorrectly: "*Coming back to A again,* `int`, *I have... (erasing on the answer sheet)... I am just writing down the numbers again as well as their corresponding position on top because I forget that they are not numbered 1, 2, 3. They are numbered 0, 1, 2, 3*". Through a critical review [12] of her initial notes, this participant was able to recognise her earlier mistake and take the necessary corrective action.

Acknowledge When Lost. While Polya [30] encourages problem solvers to discard their original plans in favour of alternative plans (if the original plans are not working), the participants in this study could not always devise an alternative. Eight of the participants (25 occurrences) acknowledged at some point during this problem-solving step that they were either lost or confused. While tracing through the second `for` loop of Q3, P8 said: *"That's the part where I am really stuck—the second loop"*, while P6 said: *"So this second `for` loop is the one that is freaking me out"*. Both participants got lost at this point because they were unable to interpret the combined index in the statement: `b[x[i]] = true;`. As another example of this strategy, P4 correctly identified option C as the answer to Q3, but when asked why he thought the answer was C, he was neither sure of himself nor of his answer. Instead, he was quick to acknowledge the limitations of his general programming skills: *"Like I said, I am not the best at these programming subjects"*. He then attempted to explain his reasoning which consequently resulted in him selecting option A (which was incorrect) as his final answer. Although this participant was able to identify the correct answer, he could not explain his reasoning. This could imply that his original selection of option C was a guess rather than a definitive answer.

4.4 Evaluate a Solution

At this stage, the participants had already identified answers to the questions. Instances of personal reflections were observed when the participants attempted to justify these answers.

Defend an Incorrect Answer. During the evaluation of their final answers, six participants (21 occurrences) remained confident while attempting to defend their incorrect answers. While P9 completely ruled out option C (the correct answer) as a possible answer to Q3 (*"so option C does not work"*), P2 remarked as follows on his answer for Q3: *"So the answer here, I think it should be 4 [option D] because this statement says, if `i` is less than `b.Length`, and `b.Length` is 5, we will increment it"*. Although P3 performed a thorough analysis and followed a reasoning strategy to answer Q6, he ended up saying: *"I believe, I am taking option A"* while option B was the correct answer. The interviewer did ask him to further explain why he thought option A was the correct option, but he still ended up giving the same incorrect answer. Although all the participants mentioned here attempted to assess or re-assess their answers, they just repeated the same flawed reasoning to arrive at the same incorrect answers. This serves as a further indication that several of the participants lacked some basic programming knowledge.

Verify Answer. While reflecting on their final answers, six participants (18 occurrences) attempted to verify or justify their selected answers while some also elaborated on their reasons for not selecting any of the other answer options. As part of the process to verify his selection of option B as the correct answer for Q6, P3 also provided justification for why he thought some of the other options were incorrect: *"Option D is totally wrong because it says if this number, in my example, 2 is greater than 3, therefore it is ascending, which is `false`. Then I have this one [pointing at the values on his written down example], so it's `true` and it returns `false`. Then it will go and take the second value, check and increment and check the other one"*. P3 followed a similar strategy while reflecting on choosing option B as the correct answer for Q3: *"Okay, this one [scribbling and looking at his written down example] will be wrong because j starts from* `zero`, *and here in this example, you don't check the numbers that are behind - you only check the numbers that are after that position. So, options A and B in this particular example will be out"*. Both of these excerpts illustrate the level of critical thinking [12] that P3 applied during the problem-solving process. These excerpts also illustrate how problem solvers can reuse some of the strategies from earlier steps (e.g., make notes, use test cases, and eliminate possibilities) while evaluating their solutions.

5 SWOT Analysis

Our next step was to perform a SWOT analysis of the SCC problem-solving strategies identified in Sect. 4 (see Table 1). Strengths refer to strategies that the participants performed well during their problem-solving processes, while the weaknesses are aspects that were lacking from their skills sets and need to be improved. The opportunities are strategies that educators can include when modelling a more ideal SCC problem-solving strategy to their students. The threats are unfavourable strategies practised by the participants that could lead to an overwhelmingly flawed problem-solving process.

Table 1. SWOT analysis of problem-solving strategies.

Strengths	Weaknesses
• Read through instructions • Identify important concepts and/or terms • Acknowledge when lost • Eliminate incorrect answer options at an early stage • Make notes for future reference • Utilise test cases • Confirm initial understanding • Recognise coding patterns • Express confidence in devised solution • Verify final answer	• Self-doubt in own programming abilities • Misinterpret basic code syntax • Fail to understand the working of basic programming structures • Reluctant to change line of thinking when lost • Struggle to interpret nested concepts • Fail to recognise all coding patterns • Incorrect use of programming terminology
Opportunities	Threats
• Realise value of devising an actual problem-solving plan • Emphasise importance of using correct programming terminology • Identify specific gaps in programming knowledge	• Disregard earlier interpretations or notes • Content to provide answers to all the questions (regardless of correctness) • Look at sections of code in isolation • Ignore "difficult" pieces of code completely • Resort to guessing when reasoning fails

6 Bottleneck Identification

By considering the identified weaknesses and threats (see Table 1), we identified six main learning bottlenecks that could prevent novice programmers from mastering problem solving in the CS discipline.

Bottleneck 1: Students Ignore Previously Understood Information at Later Stages of the Problem-Solving Process. Many of the examples illustrated in the students' excerpts (see Sect. 4) indicate the inconsistencies they exhibited while answering the given questions. It seems as if students were forgetting the correct understanding they previously had, unless they were, in the first place, not even convinced that their original understanding was correct. In this way, the students' original thinking and analysis [4] were not necessarily helping them during the entire problem-solving process.

Bottleneck 2: Students Regard Their Working Notes as a Non-crucial Component of the Problem-Solving Process. Although many of the students made conscious efforts to demonstrate how they were going about answering the questions and making their thinking logic as explicit as possible, many of the markings, comments, notes, and underlinings [18, 31, 37] they made were either overly rough or unreadable. Some of the participants also ignored their initial notes in later stages of the problem-solving process.

Bottleneck 3: Students Doubt Their Own Programming Abilities. From many of the excerpts presented in Sect. 4, it was evident that the students were not always confident of themselves. A level of doubt was observed not only with concepts they struggled to understand, but also with concepts they seemed to find fairly easy to comprehend. Consequently, they resorted to gathering fragmented pieces of information in their minds, hence making it difficult to make usable mental connections. This type of reasoning makes it difficult for one to gain a proper understanding of the question [5], and often results in guessing the answers.

Bottleneck 4: Students Are Unable to Comprehend Nested Concepts. Although all the students identified the crucial terms or concepts of computer programming, some of them were unable to fully comprehend even the most basic of these concepts and/or terms when these were combined within other programming structures. Consequently, some students were unable to organise their continuous understanding of the question around these basic concepts [8], and therefore resorted to viewing pieces of code in isolation.

Bottleneck 5: Students Are Unable to Easily Change Their Viewpoints During the Problem-Solving Process. Although some students were able to change their initial understanding, others struggled to find alternative avenues of reasoning when they got lost or stuck. Although these students attempted to make logical deductions [21], they still were unable to correctly interpret the given source code.

Bottleneck 6: Students Are More Focused on the Answer Than the Process of Arriving at the Answer. The ultimate goal of most of the students was to get to the final answer. Consequently, many of their problem-solving steps were unclear, and the expected direct reasoning [30] steps did not come to the fore. It can therefore be deduced that they regarded their final answers as more important than the actual problem-solving process. There were also instances where some students arrived at the correct answer either using flawed reasoning or displaying a lack of comprehension of the relevant programming concepts.

7 Conclusions and Future Work

The lack of problem-solving skills remains a challenge to undergraduate CS students. Comprehension of the mental moves made by students during problem solving can be essential in helping them to overcome related challenges. By focusing on Step 1 of the seven-step DtDs framework (identifying places in a course where many students consistently fail to master crucial material), this study aimed to (1) explore the problem-solving strategies utilised by novice programmers during SCC; (2) relate these strategies to Polya's four basic problem-solving steps; and (3) utilise a SWOT analysis of these strategies for the identification of problem-solving bottlenecks experienced by novice programmers. Thematic analysis of data collected by means of asking questions, observations, and artefact analysis revealed that the novice programmers in this study did not necessarily follow a well-defined problem-solving process. We were, however, able to link the specific strategies they employed to all four of Polya's basic problem-solving steps. While some of the identified strategies could be mapped to a single step, several strategies were utilised repeatedly in different stages of the problem-solving process. A SWOT analysis of the identified strategies provided further insight regarding strategies that the novices performed well (strengths), and the aspects that were lacking from their skills sets (weaknesses). We also identified several undesirable strategies (threats) that could hamper their problem-solving attempts. The listed observations point to strategies that educators could utilise to assist students in improving their problem-solving skills.

This study also illustrated a novel approach in utilising a SWOT analysis as a learning bottleneck identification tool. By considering the identified strengths and weaknesses, we formulated six bottlenecks that could prevent novice programmers from mastering problem solving in the CS discipline. The fact that these bottlenecks were identified from the problem-solving attempts of final-year undergraduate students further highlights the problem-solving difficulties that are currently not effectively addressed in introductory CS courses. Further research is needed to investigate how the identified problem-solving bottlenecks can be addressed through the application of the remaining six steps of the DtDs framework [23].

Appendix

Question 3
Consider the following source code fragment:

```
int[] x = {1, 2, 3, 3, 3};
bool[] b = new bool[x.Length];

for (int i = 0; i < b.Length; ++i)
    b[i] = false;

for (int i = 0; i < x.Length; ++i)
    b[x[i]] = true;

int count = 0;

for (int i = 0; i < b.Length; ++i)
{
if (b[i] == true)
    ++count;
}
```

After this source code is executed, count contains:

        a) 1
        b) 2
        c) 3
        d) 4
        e) 5

Question 6
The following method isSorted should return true if the array is sorted in ascending order. Otherwise, the method should return false:

```
public static bool isSorted (int[] x)
{
    //missing source code goes here
}
```

Which of the following is the missing source code from the method isSorted?

a)
```
bool b = true;
for (int i = 0; i < x.Length - 1; i++)
{
  if (x[i] > x[i + 1])
    b = false;
  else
    b = true;
}
return b;
```

b)
```
for (int i = 0; i < x.Length - 1; i++)
{
  if (x[i] > x[i + 1])
    return false;
}
return true;
```

c)
```
bool b = false;
for (int i = 0; i < x.Length - 1; i++)
{
  if (x[i] > x[i + 1])
    b = false;
}
return b;
```

d)
```
bool b = false;
for (int i = 0; i < x.Length - 1; i++)
{
  if (x[i] > x[i + 1])
    b = true;
}
return b;
```

e)
```
for (int i = 0; i < x.Length - 1; i++)
{
  if (x[i] > x[i + 1])
    return true;
}
return false;
```

References

1. Alturki, R.A.: Measuring and improving student performance in an introductory programming course. Inf. Educ. **15**(2), 183–204 (2016). https://doi.org/10.15388/infedu.2016.10
2. Anyango, J.T., Suleman, H.: Teaching programming in Kenya and South Africa: what is difficult and is it universal ? In: Proceedings of the 18th Koli Calling International Conference on Computing Education Research, pp. 1–2. ACM, New York (2018). https://doi.org/10.1145/3279720.3279744
3. Belski, I.: Teaching thinking and problem solving at university: a course on TRIZ. Creativity Innov. Manag. **18**(2), 101–108 (2009). https://doi.org/10.1111/j.1467-8691.2009.00518.x
4. Bransford, J., Brown, A., Cocking, R.: How People Learn: Brain, Mind, Experience, and School (Expanded). National Academy Press, Washington, D.C. (2000). https://doi.org/10.4135/9781483387772.n2
5. Bransford, J.D., Stein, B.S.: The Ideal Problem Solver: A Guide for Improving Thinking, Learning, and Creativity, 2nd edn. W.H. Freeman, New York (1993)
6. Braun, V., Clarke, V.: Using thematic analysis in psychology. Qual. Res. Psychol. **3**, 77–101 (2006). https://doi.org/10.1191/1478088706qp063oa
7. Cheah, C.S.: Factors contributing to the difficulties in teaching and learning of computer programming: a literature review. Contemp. Educ. Tech. **12**(2), 1–14 (2020). https://doi.org/10.30935/cedtech/8247
8. Chi, M., Glaser, R., Rees, E.: Expertise in problem solving. In: Sternberg, R. (ed.) Advances in the Psychology of Human Intelligence, vol. 1, pp. 7–75. Lawrence Erlbaum Associates Inc, New Jersey (1982)
9. Conn, C., McLean, R.: Bulletproof Problem Solving: The One Skill That Changes Everything. Wiley, Hoboken (2019)
10. Cunningham, K., Blanchard, S., Ericson, B., Guzdial, M.: Using tracing and sketching to solve programming problems: Replicating and Extending an Analysis of What Students Draw. In: Proceedings of the 2017 ACM Conference on International Computing Education Research, pp. 164–172. ACM, New York (2017). https://doi.org/10.1145/3105726.3106190
11. Diaz, A., Middendorf, J., Pace, D., Shopkow, L.: The history learning project: a department "decodes" its students. J. Am. Hist. **94**(4), 1211–1224 (2008). https://doi.org/10.2307/25095328
12. Egbert, J.: Methods of Education Technology: Principles, Practice, and Tools. Pearson, New Jersey (2017). https://opentext.wsu.edu/tchlrn445/
13. Fedorenko, E., Ivanova, A., Dhamala, R., Bers, M.U.: The Language of programming: a cognitive perspective. Trends Cogn. Sci. **23**(7), 525–528 (2019). https://doi.org/10.1016/j.tics.2019.04.010
14. Fitzgerald, S., Simon, B., Thomas, L.: Strategies that students use to trace code: an analysis based in grounded theory. In: Proceedings of the 1st International Workshop on Computing Education Research, pp. 69–80. ACM, New York (2005). https://doi.org/10.1145/1089786.1089793
15. Herrmann, J.W.: Rational decision making. In: Balakrishnan, N., Colton, T., Everitt, B., Piegorsch, W., Ruggeri, F., Teugels, J.L. (eds.) Wiley StatsRef: Statistics Reference Online, pp. 1–9. Wiley, New York (2017). https://doi.org/10.1002/9781118445112.stat07928
16. Humphrey, A.S.: SWOT analysis. Long Range Plan. **30**, 46–52 (2005)
17. Khomokhoana, P.J., Nel, L.: Decoding source code comprehension: bottlenecks experienced by senior computer science students. In: Tait, B., Kroeze, J., Gruner, S. (eds.) SACLA 2019. CCIS, vol. 1136, pp. 17–32. Springer, Cham (2020). https://doi.org/10.1007/978-3-030-35629-3_2

18. Lister, R., et al.: A multi-national study of reading and tracing skills in novice programmers. ACM SIGCSE Bull. **36**(4), 119–150 (2004). https://doi.org/10.1145/1041624.1041673
19. Loksa, D., Ko, A.J., Jernigan, W., Oleson, A., Mendez, C.J., Burnett, M.M.: Programming, problem solving, and self-awareness: effects of explicit guidance. In: Proceedings of the 2016 CHI Conference on Human Factors in Computing Systems, pp. 1449–1461. ACM, New York (2016). https://doi.org/10.1145/2858036.2858252
20. Malik, S.I., Coldwell-Neilson, J.: Impact of a new teaching and learning approach in an introductory programming course. J. Educ. Comput. Res. **55**(6), 789–819 (2017). https://doi.org/10.1177/0735633116685852
21. Malloy, C.E., Jones, M.G.: An investigation of African American students' Mathematical problem solving. J. Res. Math. Educ. **29**(2), 143–163 (1998)
22. Margulieux, L.E., Morrison, B.B., Decker, A.: Reducing withdrawal and failure rates in introductory programming with subgoal labeled worked examples. Int. J. STEM Educ. **7**(1), 1–16 (2020). https://doi.org/10.1186/s40594-020-00222-7
23. Middendorf, J.K., Pace, D.: Decoding the disciplines: a model for helping students learn disciplinary ways of thinking. New Dir. Teach. Learn. **98**, 1–12 (2004). https://doi.org/10.1002/tl.142
24. Middendorf, J., Shopkow, L.: Overcoming Student Learning Bottlenecks: Decode Your Disciplinary Critical Thinking. Stylus Publishing LLC, Sterling (2018)
25. Moore, D., Zabrucky, K., Commander, N.E.: Validation of the metacomprehension scale. Contemp. Educ. Psychol. **22**(4), 457–471 (1997). https://doi.org/10.1006/ceps.1997.0946
26. Pace, D.: The Decoding the Disciplines Paradigm: Seven Steps to Increased Student Learning. Indiana University Press, Bloomington (2017)
27. Patton, M.Q.: Qualitative Research & Evaluation Methods: Integrating Theory and Practice, 4th edn. SAGE, Thousand Oaks (2015)
28. Pinnow, E.: Decoding the disciplines: an approach to scientific thinking. Psychol. Learn. Teach. **15**(1), 94–101 (2016). https://doi.org/10.1177/1475725716637484
29. Plowright, D.: Using Mixed Methods: Frameworks for An Integrated Methodology. SAGE, Thousand Oaks (2011)
30. Polya, G.: How to Solve It: A New Aspect of Mathematical Method. Doubleday, University of Michigan (1957)
31. Powell, N., Moore, D., Gray, J., Finlay, J., Reaney, J.: Dyslexia and learning computer programming. ACM SIGCSE Bull. **36**(3), 242 (2004). https://doi.org/10.1145/1026487.1008072
32. Praveen, A.: Program comprehension and analysis. Int. J. Eng. Appl. Comput. Sci. **1**(01), 17–21 (2016). https://doi.org/10.24032/ijeacs/0101/04
33. Preece, J., Rogers, Y., Sharp, H.: Interaction Design: Beyond Human-Computer Interaction, 4th edn. Wiley, New York (2015)
34. Rahmat, M., Shahrani, S., Latih, R., Yatim, N.F.M., Zainal, N.F.A., Rahman, R.A.: Major problems in basic programming that influence student performance. Proc. Soc. Behav. Sci. **59**, 287–296 (2012). https://doi.org/10.1016/j.sbspro.2012.09.277
35. Renumol, V., Jayaprakash, S., Janakiram, D.: Classification of cognitive difficulties of students to learn computer programming. Indian Institute of Technology (2009). http://dos.iitm.ac.in/publications/LabPapers/techRep2009-01.pdf
36. Samejima, M., Shimizu, Y., Akiyoshi, M., Komoda, N.: SWOT analysis support tool for verification of business strategy. In: IEEE International Conference on Computational Cybernetics, pp. 631–635. IEEE, Talinn (2006). https://doi.org/10.1109/ICCCYB.2006.305700
37. Scalabrino, S., et al.: Improving code readability models with textual features. In: Proceedings of the 24th IEEE International Conference on Program Comprehension, pp. 1–10. IEEE, Austin (2016). https://doi.org/10.1109/ICPC.2016.7503707

38. Scherer, R., Siddiq, F., Sánchez Viveros, B.: A meta-analysis of teaching and learning computer programming: effective instructional approaches and conditions. Comput. Hum. Behav. **109**(106349), 1–18 (2020). https://doi.org/10.1016/j.chb.2020.106349
39. Shopkow, L.: How many sources do I need? Hist. Teach. **50**(2), 169–200 (2017)
40. Uwano, H., Nakamura, M., Monden, A., Matsumoto, K.I.: Analyzing individual performance of source code review using reviewers' eye movement. In: Proceedings of the 2006 Symposium on Eye Tracking Research & Applications, pp. 133–140. ACM, New York (2006). https://doi.org/10.1145/1117309.1117357
41. Yurdugül, H., Aşkar, P.: Learning programming, problem solving and gender: a longitudinal study. Proc. Soc. Behav. Sci. **83**, 605–610 (2013). https://doi.org/10.1016/j.sbspro.2013.06.115

Understanding the Significance of Enterprise Resource Planning Education in Zambia: A Case of an ERP Short Course at University of Zambia

Mampi Lubasi(✉) and Lisa F. Seymour

CITANDA, University of Cape Town, Cape Town, South Africa
mampi.lubasi@gmail.com, Lisa.seymour@uct.ac.za

Abstract. This paper investigates the significance of ERP Education on postgraduate students who took an ERP short course at the University of Zambia. The paper identifies the benefits of ERP education in Zambia as well as the contextual factors that impacted ERP education. These contextual factors were identified as industry, course and student constraints. Thematic networks showing the benefits of ERP education, the constraints on ERP education and the preliminary relationships are also presented. Universities seeking to integrate ERP systems into their curriculum and organisations seeking to hire ERP graduates would therefore benefit from these findings.

Keywords: ERP systems · ERP education · Inductive approach

1 Introduction

The integration of Enterprise Resource Planning (ERP) systems in the university curriculum were introduced to help bridge the gap in the demand for ERP expertise [1]. One of the requirements of ERP success is having employees with the appropriate skills to implement and maintain ERP systems. In this paper, the integration of ERP systems into the university curriculum is referred to as ERP education. SAP was the first vendor to offer its ERP system for classroom use to universities in 1996 as the ERP vendor believed that this would further the development of their products. The SAP alliance programme has helped universities on their programme with remote hosting, supply of classroom material and workshops for faculty [2].

It has been noted that while universities have introduced ERP education, there is limited research on the impact or added value of ERP education [1]. Therefore, calls for research have been made to determine the benefits and success of ERP education. Empirical evidence is required on whether ERP education has given ERP graduates an advantage in the workplace or not [3]. Empirical evidence is also required to determine how ERP education affects employability and salary scales [4, 5]. It has been stated that more time and resources are spent on employees without ERP education compared to the time spent on those with ERP education. Empirical evidence is therefore required to determine the value of ERP graduates to companies that hire them [2].

In this qualitative study, we investigate the impact that ERP education had on students who took an ERP course in Zambia and the contextual factors that impacted ERP education. The study seeks to answer the research question, "How does ERP education impact student capabilities and organisations and what factors impact these capabilities?" In ERP education, one size does not fit all as each context is unique. This paper, therefore, makes two contributions on the impact of ERP education. Firstly, we identify the benefits of ERP education to students in Zambia. Secondly, we identify the constraints on the impact of ERP education in Zambia. The findings would benefit academic institutions seeking to offer ERP education and organisations seeking to hire ERP graduates.

This paper is structured as follows: Sect. 1 presents the introduction and context of the study. Section 2 presents the literature review. Section 3 presents the case description. Section 4 presents the method that was used in data collection and analysis. Section 5 discusses the findings of the research. Section 6 presents the thematic networks and preliminary relationships from the findings. Section 7 presents the conclusion of the study.

2 Literature Review

Universities use five approaches in teaching ERP curriculum namely ERP training, ERP via business processes, information systems approach, ERP concepts, and hybrid approach [6]. The ERP training approach focuses on a particular ERP system and is the least preferred approach by academic institutions [6]. ERP via business processes approach focuses on business processes and uses an ERP system to enhance students' understanding of business processes and their interdependencies [6]. The information systems approach uses ERP systems to teach information system concepts [6]. The ERP concepts approach teaches general ERP systems and concepts and does not focus on one particular ERP system [6]. The hybrid approach is a hybrid of the four approaches.

Pedagogical methodologies that have been used in teaching ERP systems include hands-on experience, case teaching, technical implementation, and simulations [7]. Hands-on experience teaches navigation, exploring, processing transactions, and configuring ERP systems [7]. The hands-on experience focuses on the step-by-step execution of instructions and not on business logic [7]. Students, therefore, learn how to execute technical tasks without understanding why [7]. Hands-on experience is of benefit when integrated with class discussions [7]. Case teaching enhances process-oriented thinking and develops high order reasoning skills when integrated with hands-on experience which helps to increase student motivation and interest [7]. However, case teaching does not expose students to the challenges they will encounter in the real world with changing business processes [7]. Simulation encourages high order reasoning, decision-making skills and increases student motivation [7]. Technical implementation teaches configuration issues rather than strategic issues related to the use and adoption of ERP systems [7].

ERP education provides several benefits which include providing students with ERP skills and exposes them to ERP system concepts and business processes to enable them to get better jobs, enhancing the credibility of academic institutions in the eyes of industry, exposing students to concepts surrounding ERP adoption and implementation, and universities can use ERP education as a marketing tool to attract potential students [8]. ERP education has not been an easy task for universities as they have encountered several challenges. Some of the challenges they have faced are inadequate ERP skills among academic staff, lack of appropriate material to incorporate into curriculum, selecting an appropriate enterprise system, high costs of procurement and maintenance of ERP systems, lack of vendor support, difficulty in achieving learning outcomes, and creating a curriculum that has both business-centric and technology-oriented approaches [9].

A study on the integration of an SAP Financial module into a university Accounting curriculum showed that students felt that ERP training had the potential to enhance their careers and help them find more job opportunities [10]. Another study found that ERP training and user satisfaction influences individual impact or performance [11]. To develop students' professional skills, university collaboration with industry is important, enabling student internships and mentoring from potential employers [12].

Many of the studies on ERP education have been conducted in developed countries and the findings may not be valid in developing countries. Therefore, much of the research suffers from a generalizability problem [13]. In Africa, many universities have poorly developed and distributed technology infrastructure with limited internet access, bandwidth, and unreliable access to electricity [14]. Hence there is a need to understand the impact of ERP education in developing countries and the contextual factors that impact ERP education. Understanding the impact of ERP education on students and their performance in the industry would assist in justifying investment in ERP education [13].

3 Case Description

In Sub-Saharan Africa, the Enterprise Systems Education for Africa (ESEFA) programme has partnered with universities to teach ERP fundamentals using SAP ERP [15]. ERP education was introduced at the University of Zambia (UNZA) in October 2014 through the ESEFA programme. The ERP course was offered to Computer Science undergraduate students and postgraduate students in Engineering, Economics and Computer Science. The postgraduate students from the different disciplines all attended the course together as one class. The ERP course content was offered to postgraduate students as a short course and ran for about six weeks. About 41 postgraduate students took the ERP course.

The course content used SAP ERP and had a case study component based on a Zambian bamboo bike producing company (Zambikes), labs and lectures. The following topics were covered in the ERP course: Business Processes, Enterprise Systems Knowledge, SAP ERP Navigation and Reporting, Procure to Pay, Sales to Cash, and Inventory Management. The Zambikes video was shown at the beginning of the course and presented the organisation's business processes. Most of the students who took the course were from Computer Science and Engineering with an IT background. Very few students from Economics took the ERP course. There was an exam at the end of the course and an SAP University Alliances certificate of proficiency was issued upon successfully passing the exam. The ERP course at UNZA is an example of the ERP via business processes approach and the Hands-on experience method.

4 Method

The research paradigm or philosophy that was used is interpretivism. Interpretivism adopts the position that our knowledge of reality is a social construction by human actors [16]. Interpretivism was used because it has the potential to produce deep insights in Information Systems research as it seeks to understand a phenomenon of study through the meanings assigned by human actors [17]. An interpretive approach helped to provide an understanding of the Zambian context and how it impacted ERP education. The research approach or reasoning that was used is induction. An inductive approach was used because it allows research findings to emerge from dominant or significant themes inherent in raw data without the restraints imposed by a theoretical framework [18]. Thomas' general inductive approach was used. An inductive approach can assist in condensing raw data into brief or summary format, establish clear links between research objectives and summary findings derived from raw data, and develop a framework of the underlying structure of experiences or processes that are evident in the raw data [18]. Thomas' general inductive approach "provides an easily used and systematic set of procedures for analyzing qualitative data that can produce reliable and valid findings [18]."

Purposive sampling was used in selecting participants for the interviews. Purposeful sampling was used because it selects the most productive sample that will aid in answering the research question [19]. Data collection was cross-sectional as it was collected at one point in time. Data was collected from twenty-one postgraduate students at UNZA using semi-structured interviews in November 2016 and in February 2019. During the second round of interviews in 2019, more students from Engineering were interviewed as only one Engineering student was interviewed in the first round of interviews in 2016. Semi-structured interviews were used as they provide flexibility and enable the development of responses from the interviewee. The availability of the students at the time of the interviews was the determining factor when selecting interviewees. An initial set of questions was prepared for the first round of interviews. The questions were adjusted for the second round of interviews based on the findings from the first round of interviews. Table 1 shows the participant information.

Table 1. Participant information

Programme of study	No. of respondents	Code
Economics/Finance	2	ACC
Computer Science	11	CS
Engineering	8	ENG

The themes that emerged were coded in NVivo software. The following steps were followed: multiple readings of the raw data guided by the evaluation objectives, development of categories or themes from the raw data into a model or framework and determining relationships with other categories or themes [18].

5 Findings and Discussion

This section discusses the findings from the study. The themes that emerged from the study are presented in the tables and the themes are discussed.

5.1 Benefits of ERP Education to Students

ERP Education provided several benefits to postgraduate students. The themes that emerged are shown in Table 2 and are discussed in this section.

Table 2. Benefits of ERP education

Benefits of ERP education	No. of respondents	No. of statements
Increases self-confidence in using ERP systems	16	16
Increases employment prospects	15	22
Provides hands-on experience with ERP systems	11	11
Increases understanding of how a business is run	10	11
Increases likelihood of getting a higher salary	8	8
Increases decision-making on ERP software acquisition	4	4
Enables further training	3	3
Increases ability to assist ERP system users in solving ERP related problems	2	3
Provides insight on current software limitations	1	2

Increases Self-confidence in Using ERP Systems. As a result of going through the ERP course, sixteen respondents said that the course increased their self-confidence in using SAP ERP systems. Literature shows that students who took an ERP course in Canada expressed confidence in their ability to manage ERP systems although the students viewed themselves as potential users of ERP systems and not managers [20]. The results in this study are similar to the results in literature showing that ERP education increases students' confidence in using ERP systems.

> "I can confidently say I know how to use SAP even if there is a shortage of manpower at the office, I can use SAP." (ENG5)

Increases Employment Prospects. Fifteen respondents felt that the ERP course increased their employment opportunities and marketability as they now have an additional skill. Literature shows that students who have gone through ERP courses have increased employment prospects and increased market value [2, 20]. The results in this study are similar to those in literature, however, most of the students in this research are still working for the same organisations they worked for before taking the ERP course.

"Recently, I tried to apply for a job and told them that I had done that, they called me, though I turned it down." (ENG6)

Provides Hands-on Experience with ERP Systems. Eleven respondents said that the course provided hands-on experience with ERP systems through the labs that were part of the course. Literature shows that by completing hands-on ERP activities with SAP, students can better understand ERP systems and develop SAP skills [21]. Results in this study and in literature are similar in that hands-on exercises or labs help students have experience with ERP systems.

"Since it was practical, at least even if I work in an organisation where they use ERP, I'll be able to use that because the labs gave me hands-on experience with the ERP system." (ACC2)

Increases Understanding of How a Business Is Run. Ten respondents felt that the course had increased their understanding of how a business is run through the different business processes they went through in the course and their interdependencies. These findings are similar to other studies in literature that show that students who went through an ERP course had increased knowledge and understanding of business processes and functions [22, 23].

"It gave me a view of what goes on in running of companies and also what it would take to start up your own and what would be needed of you in that case." (CS8)

Increases Likelihood of Getting a Higher Salary. Eight respondents felt that the course increased their likelihood of getting a higher salary if they moved to different organisations that require ERP skills. In their current organisations, the respondents have received no salary increment despite the ERP qualification they have. Literature shows that students with an intensive ERP background had higher starting salaries [2]. Another study [24] shows that students who went through an ERP course based on an ERP simulation game had higher starting salaries. The results in this study show that students did not get higher salaries, but they felt that the course had increased their likelihood of getting higher salaries if they obtained jobs requiring ERP skills, while literature shows that ERP graduates had higher starting salaries after going through an ERP course.

"Most of the multi-nationals operating here I think all rely on ERP systems so I think it would definitely lead to a higher salary." (CS8)

Increases Decision-Making on ERP Software Acquisition. Four respondents felt that the ERP course increased their decision-making on ERP software acquisition in the workplace. Literature shows that students who went through an ERP course based on a simulation game had increased decision-making skills [25]. The decision-making in this research was on ERP software acquisition in their organisations while the decision-making in literature was on business decisions made during an ERP simulation game that was part of an ERP course.

"In terms of making decisions for acquiring software for the company I think that has changed very much because of doing the course." (ENG1)

Enables Further Training. Three respondents felt that the ERP course laid a foundation for them to pursue further training in similar software packages and in finance. Since ERP education introduces students to business processes and concepts [23] and provides hands-on experience with ERP systems [21], it, therefore, lays a good foundation for a student's future experience in the industry.

"In fact, after going through that course it motivated me to do a business course. I went on to do a business course and it made more sense because I've the ERP." (ENG5)

Increases Ability to Assist ERP System Users in Solving ERP-Related Problems. Two respondents felt that the ERP course increased their problem-solving skills as they are now able to assist other users in resolving ERP related problems in the workplace. It has been observed in other studies that students who have gone through ERP courses based on an ERP simulation game develop problem-solving skills [26].

"We have a pool office where I work from and most of the times people call you for troubleshooting, when they have problems, they ask you what you think this would be and that has become common place presently because of the knowledge I have acquired so yeah that has really given me that enhanced feeling of being able to handle problems." (ACC1)

Provides Insight on Current Software Limitations in the Organisation. One respondent felt that the ERP course had provided insights on the limitations of the software applications they currently have in their organisation.

"You see the limitations of the current applications that you are running. So, when you see those limitations then it means that you are appreciating the course that you've done then you see that if we used this then we can be able to do more." (ENG1)

5.2 Industry Constraints

Several industry constraints on the impact of ERP Education in Zambia were identified. These constraints are presented in Table 3 and are discussed in this section.

Lack of Awareness of ERP Education at University. Nine respondents said that organisations in Zambia are not aware of ERP education at UNZA hence the organisations do not know about the presence of ERP graduates in the country. Literature shows that when there is industry participation in ERP education, organisations will have more knowledge about the skills of university students, and they may also provide employment opportunities to university graduates [27]. Therefore, where there is no collaboration between industry and academia, organisations may not be aware of certain courses offered at universities.

Table 3. Industry constraints

Industry constraints	No. of respondents	No. of statements
Lack of awareness of ERP education at university	9	10
ERP systems are not implemented in workplace	8	8
Dependence on foreign expertise or consultants	8	11
Lack of understanding of ERP systems	3	4
No job adverts requiring ERP skills	2	2
ERP qualification is not appreciated in workplace	2	3

"I think most people are not aware that this programme is offered locally, and we have got graduate students from University of Zambia who have been trained there and have got our own local trainers." (ENG6)

ERP Systems Are Not Implemented in Workplace. Eight respondents said that the knowledge gained through the ERP course did not have much impact as the knowledge is not being applied in the workplace. The respondents mentioned that they do not have ERP systems implemented in the workplace. Literature shows that high implementation and maintenance costs are one of the reasons ERP systems are not adopted by organisations [28]. In developing countries, ERP is still in its early stages because of limited IT infrastructure, low IT maturity, government policies and lack of ERP experience [28].

"We don't have maybe at the moment as an institution capacity to afford such licenses to deploy such a solution in our setting, so I think that is one of the limiting factors which means there is nowhere we can continuously apply the knowledge that was learnt." (CS10)

Dependence on Foreign Expertise or Consultants. Eight respondents felt that organisations in Zambia prefer foreign expertise or consultants to local expertise thus disadvantaging local ERP graduates. ERP systems in Zambia are implemented and maintained by foreign consultants hence organisations continue to rely on them. Literature shows that since ERP implementation requires product-specific and business skills the dependence on consultants as implementation partners greatly increases [29]. Organisations, therefore, outsource ERP skills rather than investing and developing them internally [29].

"I think most organisations still prefer to look outside the country for expertise especially in ERP which isn't that big locally, so I think the decision for them to look out rather than look in, I think that's a big hindrance for most of the graduates of the ESEFA programme locally." (CS8)

Lack of Understanding of ERP Systems. Three respondents felt that organisations do not understand ERP systems hence they do not understand the value that ERP graduates could add. Literature shows that IT maturity levels are low in some firms in developing countries [28], hence this could lead to some organisations not understanding ERP systems.

> *"Not everybody understands the work which is required to somehow have a fully operational enterprise system because everyone forgets the aspect of putting in the raw data to fit the components." (CS9)*

No Job Adverts Requiring ERP Skills. Two respondents said that they had not seen any job advertisements requiring ERP skills hence they could not apply for jobs using the ERP qualification. Literature shows that many organisations rely on consultants for ERP implementation and maintenance [29], hence this could lead to many organisations in Zambia not advertising ERP jobs.

> *"I haven't seen adverts maybe I think when you're already in an organisation and then they want to implement it then you would have that opportunity, but I haven't seen organisations that are going to advertise and say they want someone with this skill." (CS2)*

ERP Qualification Is Not Appreciated in Workplace. Two respondents felt that the ERP qualification was not valued or recognized in the workplace compared to other qualifications they were pursuing hence this was a restriction. Literature shows that some students who took an ERP course were not sure if they gained skills that would be rewarded in the job market [20]. Other studies however show that the ERP qualification is valued in the workplace as ERP graduates had to start higher salaries [2, 24]. Literature shows that organisations value ERP skills while in this research the students felt that ERP skills were not valued.

> *"I have the knowledge and they benefit from it but then my paper is not as valued as the other qualifications that we are busy pursuing, so I think to begin with, the organisation has to appreciate and value the training that we have." (ACC1)*

5.3 Course Constraints

A number of course constraints were also identified. The constraints are shown in Table 4 and are discussed in this section.

ERP Course Content Was Too Basic. Three respondents felt that the ERP course content was too basic for them to be able to compete with consultants hence they needed a more advanced course. Literature shows that students who took an ERP course felt that they did not get enough technical practice and training from the course [30]. The results in this study and in literature are similar in that the students felt that they did not get enough training and practice.

> *"So, the fact that the course is also basic they need to like maybe teach something more enhanced so that our students are able to compete with the consultants." (ENG3)*

Table 4. Course constraints

Course constraints	No. of respondents	No. of statements
ERP course content was too basic	3	5
Lack of university collaboration with industry	3	3
Lack of integration of ERP course with university qualifications	3	3
Connectivity and setup challenges	2	2
SAP interface was difficult to navigate	1	1
SAP Tutorial software was restricted to university	1	1

Lack of University Collaboration with Industry. Three respondents felt that university collaboration with industry would enable organisations to be involved in the ERP programme at UNZA thus providing opportunities for students. Literature shows that collaboration between universities and industry helps to bridge the gap between the skills of university graduates and those required by industry [27]. It also helps to build stronger links between universities and industry thus helping to motivate students and providing employment to students [27]. In this research, it was noted that there was a lack of collaboration between the university and industry while literature shows how universities can benefit when there is a collaboration with industry.

> *"More interaction with industry because for some of the people that do it like the under graduates, they haven't been in industry, so they still are looking at it from just a theoretical part of it." (ENG1)*

Lack of Integration of ERP Course with University Qualifications. Three respondents felt that the ERP course needs to be combined or integrated with another qualification to be beneficial to an organisation. On its own, it may not add value to the employing organisation or ERP graduate. Hence, the respondents felt that an academic qualification or other professional qualification was necessary with ERP training as an added advantage.

> *"The ERP course that I have, I don't think it's a standalone knowledge like someone would just want to have that and then you become employable. I think you need to have it in conjunction with another course so that you become more beneficial to the organisation that employs you." (ACC1)*

Connectivity and Setup Challenges. Two respondents said that at times there were connectivity issues during the lessons. Literature shows that most universities in Sub-Saharan Africa have limited internet access, bandwidth, and unreliable access to electricity [14] which can lead to connectivity challenges. Another study shows that students who took an ERP course based on SAP at a college in Saudi Arabia encountered account setup and connectivity challenges [31].

"Yeah, the learning experience was wonderful although in the beginning we had some hitches getting to start and running of course, connecting to the system, getting the passwords and the usernames I think after we had gone through that stage then we were able to get through." (ENG1)

SAP Interface Was Difficult to Navigate. One respondent felt that the SAP ERP interface was not user friendly hence it was difficult to navigate. Literature shows that students who took a similar ERP course in Nigeria felt that the SAP tutorial software should be more user friendly to facilitate proficiency [32]. In another study, students also found the SAP screens and interface challenging to navigate [30].

"I think the only challenge was the interface of the application. I think getting people to work with it regularly would be very, very difficult." (CS3)

SAP Tutorial Software Was Restricted to University. One respondent felt that since the software could only be accessed from the university, this had restricted her from practicing at home or elsewhere. Literature shows that students in Nigeria who took a similar ERP course also identified this as a constraint [32]. For a user to be able to access SAP, the SAP GUI needs to be installed on the computer. The IP address of the computer should also be allowed to access the SAP server as there are IP restrictions on accessing the SAP server.

"The only thing is, the software is confined to here, so you cannot really practice at home. You just have to practice when you're here, like it's a software that is not shared." (CS6)

5.4 Student Constraints

The student constraints identified are presented in Table 5 and are discussed in this section.

Table 5. Student constraints

Student constraints	No. of respondents	No. of statements
Inadequate prior knowledge	7	7
Working on ERP systems is not in line of work	5	5
Limited time to devote to course	4	6
Unable to obtain certificate	2	2

Inadequate Prior Knowledge. Six respondents from an IT background felt that Accounting principles were a challenge in the course as they had no accounting background. One respondent from an accounting background felt that IT jargon was a challenge in the course. Hence some students faced challenges due to inadequate prior knowledge. Literature shows that inadequate knowledge of accounting and finance was a challenge for students from an IT background who took an ERP course in Australia [30]. In another study [25], it was found that students need to have a basic understanding of the business before attempting to learn business processes from ERP systems. The results in this study and in literature show that when students have inadequate prior knowledge, they may experience challenges in understanding certain concepts taught in ERP courses.

> *"Those were actually the biggest challenges that we also had on the credit and debit. We are not coming from that background, so we had a serious challenge." (CS1)*

Working on ERP Systems is Not in Line of Work. Five respondents mentioned that working on ERP systems is not in their line of work hence this had restricted them from applying their ERP knowledge.

> *"Basically, it's because it's not really in my line of work. There are other people that use those systems." (ENG4)*

Limited Time to Devote to Course. Four respondents said that they had limited time to dedicate to the ERP course as they were working fulltime and working on the main programme of study at UNZA. Literature shows that students who took an ERP course felt that the course presented too much work and the pace was too fast hence they requested more time in the lab [20]. In another study [32], students felt that the time allocated to the labs was too short. In this study, the students did not have the time to dedicate to the ERP course due to other commitments while literature shows that students needed more time on the course.

> *"So, the only issue that I had is the fact that my time at that particular point was limited because I was working full time and also doing the Masters programme full time so that really pushed me a little bit." (ENG4)*

Unable to Obtain Certificate. Two respondents felt that since they did not manage to obtain ERP certificates due to either not passing or writing the exam, this had restricted them when applying for other jobs as they had no certificates to prove that they did the ESEFA ERP course. Literature shows that students' perceived value of a vendor certificate such as SAP ERP will motivate them to pursue that certificate to improve their job prospects. [33].

> *"I didn't even get the certificate therefore it is even very difficult for me to talk about it when I've been called for interviews." (ENG2)*

6 Thematic Networks

The themes that emerged from the research are presented in the thematic networks. Figure 1 shows a thematic network on the benefits of ERP Education. The dominant themes on the benefits of ERP education were increased self-confidence in using ERP systems, increases employment prospects, provides hands-on experience with ERP systems, increases understanding of how a business is run, increased likelihood of getting a higher salary increases decision-making on ERP software acquisition, and enables further training. These themes were considered dominant as they were cited by three or more respondents and are presented in bold font in the thematic network.

Some preliminary relationships between benefits were also identified from the analysis of the data as also shown in Fig. 1. The solid lines represent the thematic network and the relationships between themes are represented by the dotted lines. Hands-on experience with ERP systems led to increased self-confidence in using ERP systems. Increased self-confidence in using ERP systems led to increased ability to assist ERP system users in solving ERP related problems, increased ability to assist ERP system users in solving ERP related problems, in turn, led to increased self-confidence in using ERP systems. Further training led to increased decision-making on software acquisition in the organisation.

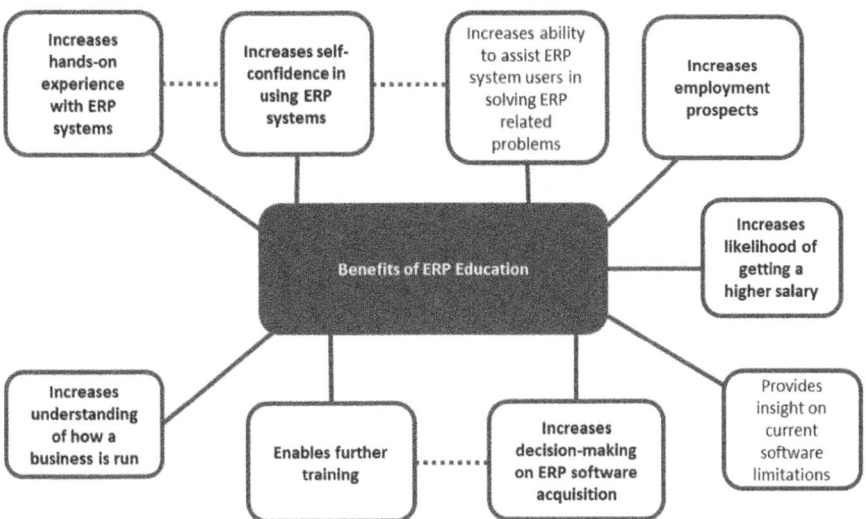

Fig. 1. Benefits of ERP education thematic network

Figure 2 shows a thematic network on the constraints on ERP education. The dominant industry constraints were lack of awareness of ERP education at university, ERP systems are not implemented in workplace, dependence on foreign expertise or consultants, and lack of understanding of ERP systems. The dominant course constraints were ERP course content was too basic, lack of university collaboration with industry, and lack of integration of ERP course with university qualifications. The dominant student constraints were inadequate prior knowledge, working on ERP systems is not in the line of work, and limited time to devote to ERP course. These themes were considered dominant as they were cited by three or more respondents and are presented in bold font in the thematic network.

Some preliminary relationships between constraints were identified from the analysis of the data as also shown in Fig. 2. The solid lines represent the thematic network and the relationships between themes are represented by the dotted lines. Connectivity and setup challenges reduced the limited time to devote to the course. Since some students had limited time to devote to the ERP course, they were then unable to obtain certificates. Due to a lack of university collaboration with industry, there was a lack of awareness of ERP education at university. Dependence on foreign expertise or consultants by organisations was impacted by the fact that the knowledge from the ERP course was too basic. The preliminary relationships show that course constraints increase student constraints and prevent a reduction of industry constraints.

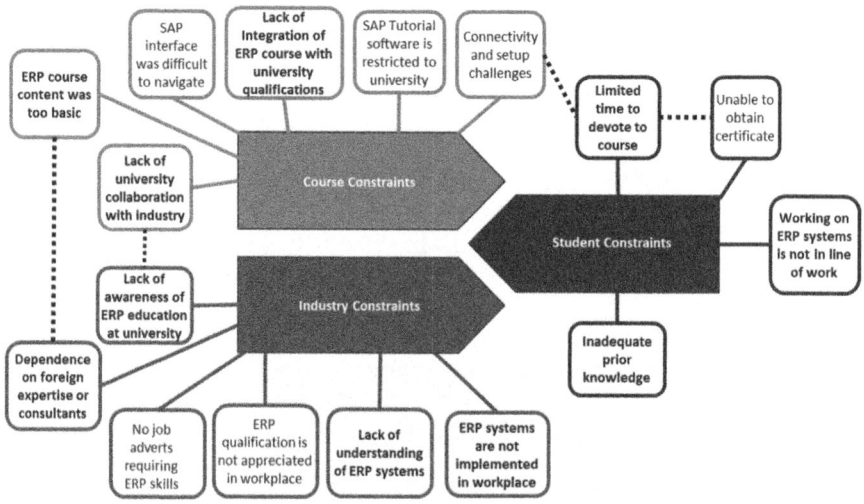

Fig. 2. Constraints on ERP education thematic network

7 Conclusion

The aim of this paper was to investigate the benefits of ERP education on postgraduate students who took an ERP short course at UNZA and identify the constraints that impacted ERP education using a general inductive approach. This research found that the dominant benefits of ERP education were increases self-confidence in using ERP systems, increases employment prospects, provides hands-on experience with ERP systems, increases understanding of how a business is run, increases likelihood of getting a higher salary, increases decision-making on ERP software acquisition in the organisation, and enables further training. The contextual factors that impacted ERP education were identified as industry, course and student constraints. The results in this study show that context should be taken into consideration when introducing ERP courses into the university curriculum as it has an impact on ERP education. The results in this paper would therefore be useful for other universities seeking to integrate ERP systems into their university curriculum and organisations seeking to hire ERP graduates.

A limitation of the study was that very few students from Economics took the ERP course. Most of the students who took the course were from Computer Science and Engineering with an IT background. Future research is required to investigate the industry's perspective on ERP education in Zambia.

References

1. Ravesteyn, P., Kohler, A.: Industry participation in educating enterprise resource planning. Commun. IIMA **9**(2), 45–55 (2009)
2. Sager, J., Mensching, J., Corbitt, G., Connolly, J.: Market power of ERP education – an investigative analysis. J. Inf. Syst. Educ. **17**(2), 151–161 (2006)
3. Bradford, M., Vijayaraman, B.S., Chandra, A.: The status of ERP integration in business school curricula: results of a survey of business schools. Commun. Assoc. Inf. Syst. **12**, 437–456 (2003)
4. Hepner, M., Dickson, W.: The value of ERP curriculum integration: perspectives from the research. J. Inf. Syst. Educ. **24**(4), 309–326 (2013)
5. Fedorowicz, J., Gelinas Jr, U.J., Usof, C., Hachey, G.: Twelve tips for successfully integrating enterprise systems across the curriculum. J. Inf. Syst. Educ. **15**(3), 235–244 (2004)
6. Hawking, P., McCarthy, B., Stein, A.: Second wave ERP education. J. Inf. Syst. Educ. **15**(3), 327–332 (2004)
7. Wang, M., Hwang, D.: An innovative framework of integrating ERP into IS 2010 model curriculum. Commun. IIMA **11**(3), 75–86 (2011)
8. Bradford, M., Vijayaraman, B.S., Chandra, A.: The status of ERP integration in business school curricula: results of a survey of business schools. Commun. Assoc. Inf. Syst. **12**(1), 26 (2003)
9. Akre, V.L., Rajan, A., Nasseri, N.: Enterprise systems (ES) integration into academic curriculum across multiple campuses of a leading academic institution in the UAE. In: International Conference on Current Trends in Information Technology (CTIT), pp. 90–95 (2013)
10. Saidi, F., Abdulkarim, M.E., Ousama, A.A.: Factors affecting the integration of the SAP-financial accounting module into an accounting curriculum: evidence from a gulf-based university. Int. J. Smart Technol. Learn. **1**(3), 218–243 (2019)
11. Costa, C.J., Aparicio, M., Raposo, J.: Determinants of the management learning performance in ERP context. Heliyon **6**(4), e03689 (2020)
12. Qiu, M., Xu, Y., Omojokun, E.O.: To close the skills gap, technology and higher-order thinking skills must go hand in hand. J. Int. Technol. Inf. Manage. **29**(1), 98–123 (2020)
13. Garavan, T., et al.: Measuring the organizational impact of training: the need for greater methodological rigor. Hum. Resour. Dev. Q. **30**(3), 291–309 (2019)
14. Mahanga, K.M., Seymour, L.F.: Enterprise resource planning teaching challenges faced by lecturers in Africa. In: Proceedings of the 9th IDIA Conference, IDIA2015, Nungwi, Zanzibar, pp. 343–353 (2015)
15. ESEFA. http://www.esefa.uct.ac.za/esefa/about/the-programme. Accessed 21 Aug 2021
16. Walsham, G.: The emergence of interpretivism in IS research. Inf. Syst. Res. **6**(4), 376–394 (1995)
17. Klein, H.K., Myers, M.D.: A set of principles for conducting and evaluating interpretive field studies in information systems. MIS Q **23**, 67–93 (1999)
18. Thomas, D.R.: A general inductive approach for analyzing qualitative evaluation data. Am. J. Eval. **27**(2), 237–246 (2006)
19. Marshall, M.N.: Sampling for qualitative research. Fam. Pract. **13**(6), 522–526 (1996)

20. Davis, C.H., Comeau, J.: Enterprise integration in business education: design and outcomes of a capstone ERP-based undergraduate e-business management course. J. Inf. Syst. Educ. **15**(3), 287–299 (2004)
21. Pridmore, J., Deng, J., Turner, D., Prince, B.: Enhancing student learning of ERP and business process knowledge through hands-on ERP exercises in an introductory management of information systems course. In: Southern Association for Information Systems Conference (SAIS), vol. 31 (2014)
22. Fulford, R.: Effective education using information systems as cognitive tools. Proc. Microsoft Dyn. Acad. Preconf. **2**(1), 8–14 (2011)
23. Seethamraju, R.: Enterprise systems (ES) software in business school curriculum-evaluation of design and delivery. J. Inf. Syst. Educ. **18**(1), 69–83 (2007)
24. Cronan, T.P., Douglas, D.E.: A student ERP simulation game: a longitudinal study. J. Comput. Inf. Syst. **53**(1), 3–13 (2012)
25. Monk, E.F., Lycett, M.: Measuring business process learning with enterprise resource planning systems to improve the value of education. Educ. Inf. Technol. **21**(4), 747–768 (2014). https://doi.org/10.1007/s10639-014-9352-6
26. Foster, S., Hopkins, J.: ERP simulation game: establishing engagement, collaboration and learning. In: PACIS, vol. 62 (2011)
27. Hawking, P., McCarthy, B.: Industry collaboration: a practical approach for ERP education. In: Proceedings of the Australasian Conference on Computing Education, pp. 129–133 (2000)
28. Abdellatif, H.J.: ERP in higher education: a deeper look on developing countries. In: 2014 International Conference on Education Technologies and Computers (ICETC), pp. 73–78. IEEE (2014)
29. Kumar, V., Maheshwari, B., Kumar, U.: An investigation of critical management issues in ERP implementation: empirical evidence from Canadian organizations. Technovation **23**(10), 793–807 (2003)
30. Seethamraju, R.: Enhancing student learning of enterprise integration and business process orientation through an ERP business simulation game. J. Inf. Syst. Educ. **22**(1), 19–29 (2011)
31. Shanneb, A.: Incorporating SAP® ERP training into industrial college education: a usability evaluation. I. J. Educ. Manage. Eng. **5**, 1–9 (2020)
32. Alawode, O.O., Anyaeche, C.O., Osunade, O.: Perception and evaluation of enterprise systems education in University of Ibadan, Nigeria. In: Proceedings of the 9th IDIA Conference, IDIA2015, Nungwi, Zanzibar, pp. 434–450 (2015)
33. Kung, H.-J., Kung, L.: Pursuing a vendor-endorsed ERP award for better job prospect: students' perceptions. Inf. Syst. Educ. J. **15**(3), 29–41 (2017)

Author Index

Bradshaw, Karen 85
Broekman, Andrew 27

Calitz, Andre P. 3

Dick, Geoffrey 43

Janse van Rensburg, J. T. 67

Khomokhoana, Pakiso J. 132

le Roux, Daniel B. 118
Liebenberg, Janet 51
Lubasi, Mampi 149

Marshall, Linda 27
Mase, Mokotsolane Ben 102
Milne, Shannon 85

Nel, Liezel 102, 132

Seymour, Lisa F. 149

van der Linde, Suné 51

The manufacturer's authorised representative in the EU is Springer Nature Customer Service Centre GmbH, Europaplatz 3, 69115 Heidelberg, Germany. If you have any concerns regarding our products, please contact ProductSafety@springernature.com

Printed and bound by CPI Group (UK) Ltd, Croydon, CR0 4YY
25/03/2026
02078181-0015